Timo Heister, Leo G. Rebholz
Introduction to Scientific Computing
De Gruyter Textbook

Also of Interest

Computational Technologies. A First Course
Vabishchevich (Ed.), 2015
ISBN 978-3-11-035992-3, e-ISBN (PDF) 978-3-11-035995-4,
e-ISBN (EPUB) 978-3-11-039103-9

Computational Technologies. Advanced Topics
Vabishchevich (Ed.), 2015
ISBN 978-3-11-035994-7, e-ISBN (PDF) 978-3-11-035996-1,
e-ISBN (EPUB) 978-3-11-038688-2

Computer Arithmetic and Validity
Kulisch, 2013
ISBN 978-3-11-030173-1, e-ISBN (PDF) 978-3-11-030179-3

Computer Simulation in Physics and Engineering
Steinhauser, 2012
ISBN 978-3-11-025590-4, e-ISBN (PDF) 978-3-11-025606-2

Computational Methods in Applied Mathematics
Carsten Carstensen (Editor-in-Chief)
ISSN 1609-4840, e-ISSN 2083-9389

www.degruyter.com

Timo Heister, Leo G. Rebholz

Introduction to Scientific Computing

For Scientists and Engineers

DE GRUYTER

Mathematics Subject Classification 2010
65F05, 65F15, 65F35, 65G50, 65L05, 65L10, 65L12, 65L20, 65Y20

Authors
Prof. Timo Heister, PhD
Clemson University
Mathematical Sciences
Clemson SC 29634
USA
heister@clemson.edu

Prof. Leo G. Rebholz, PhD
Clemson University
Mathematical Sciences
Clemson SC 29634
USA
rebholz@clemson.edu

ISBN 978-3-11-035940-4
e-ISBN (PDF) 978-3-11-035942-8
e-ISBN (EPUB) 978-3-11-038680-6

Library of Congress Cataloging-in-Publication Data
A CIP catalog record for this book has been applied for at the Library of Congress.

Bibliographic information published by the Deutsche Nationalbibliothek
The Deutsche Nationalbibliothek lists this publication in the Deutsche Nationalbibliografie;
detailed bibliographic data are available on the Internet at http://dnb.dnb.de.

© 2015 Walter de Gruyter GmbH, Berlin/Boston
Typesetting: PTP-Berlin, Protago TEX-Production GmbH
Printing and binding: CPI books GmbH, Leck
♾Printed on acid-free paper

Printed in Germany

www.degruyter.com

Author LR dedicates this book to Laina and LJ.

Author TH dedicates this book to his parents and siblings (who might never see this) and to the Clemson students studying with this book.

Preface

This book is intended for sophomore-level engineering and science students interested in numerical methods for solving problems such as differential equations, eigenvalue problems, linear systems, etc., and could be appropriate for mathematics majors as well, if the instructor supplements this text with more proofs. It has evolved from our lecture notes during the many times we have taught the course since 2010.

We assume a knowledge of Calculus through multivariable Calculus (typically called Calculus III in the United States), and we teach the course at Clemson with a first course in Ordinary Differential Equations being a co-requisite. Also, we assume students have at least seen and used MATLAB in some minimal capacity.

We decided to write this book because we wanted the text for this course to have the following:
1. Low cost. Many books that cover similar material cost upwards of $100 or even $150, and we refuse to be part of that. Hence we have chosen a publisher that provides a much lower cost.
2. Simple programming examples. This is an introductory programming course, and students need to learn the basics before all the bells and whistles. Hence whenever possible, we give code that is as simple as possible, without overcomplicating the main ideas and without excessive commenting.
3. Cover the fundamentals. For any given problem for which we use numerical methods, there are often dozens of approaches. We believe an introductory text should focus on the basics, and so we give just the most common approaches, and discuss them in more detail.

As this book is a first edition, there are bound to be remaining typos and mistakes, despite our best efforts. We would greatly appreciate any users of this book to point them out to us. We would also appreciate any other constructive criticism regarding the presentation of the material.

Software

Algorithms given in the text are written in the language of MATLAB and Octave. Currently at Clemson, all students have free access to MATLAB. Octave is a free version of MATLAB, which has almost all of the same functionality. Newer versions of MATLAB have more bells and whistles, but for the purposes of this book, either MATLAB or Octave can be used.

We have created a website for the codes used in this book, where all MATLAB/Octave codes from the text can be downloaded. We have also posted Python versions of the codes: http://www.math.clemson.edu/~heister/scicompbook/

Acknowledgement: We wish to thank Mine Akbas, Abigail Bowers, Chris Cox, Keith Galvin, Evelyn Lunasin, and Ashwin Trikuta Srinath for their help in the preparation of this manuscript.

The author LR thanks the National Science Foundation for partial support of his research, and indirectly in the writing of this book, through the grant DMS1112593.

The author TH is supported in part through the Computational Infrastructure in Geodynamics initiative (CIG), through the National Science Foundation under Award No. EAR-0949446 and The University of California – Davis.

Contents

1 Introduction

1.1 Why study numerical methods?

A fundamental questions one should ask before spending time and effort learning a new subject is "Why should I bother?". Aside from the fun of learning something new, the need for this course arises from the fact that most mathematics done in practice (therefore by engineers and scientists) is now done on a computer. For example, it is common in engineering to need to solve more than 1 million linear equations simultaneously, and even though we know how to do this 'by hand' with Gaussian elimination, computers can be used to reduce calculation time from years (if you tried to do it by hand – but you would probably make a mistake!) to minutes or even seconds. Furthermore, since a computer has a finite number system and each operation requires a physical change in the computer system, the idea of having infinite processes such as limits (and therefore derivatives and integrals) or summing infinite series (which occur in calculating sin, cos, and exponential functions for example) cannot be performed on a computer. However, we still need to be able to calculate these important quantities, and thus we need to be able to approximate these processes and functions. Often in scientific computing, there are obvious ways to do approximations; however it is usually the case that the obvious ways are not the best ways. This raises some fundamental questions:

- How do we best approximate these important mathematical processes/operations?
- How accurate are our approximations?
- How efficient are our approximations?

It should be no surprise that we want to quantify accuracy as much as possible. Moreover, when the method fails, we want to know why it fails. In this course, we will see how to rigorously analyze the accuracy of several numerical methods. Concerning efficiency, we can never have the answer fast enough[1], but often there is a trade-off between speed and accuracy. Hence we also analyze efficiency, so that we can 'choose wisely'[2] when selecting an algorithm.

Thus, to put it succinctly, the purpose of this course is to

- Introduce students to some basic numerical methods for using mathematical methods on the computer.
- Analyze these methods for accuracy and efficiency.
- Implement these methods and use them to solve problems.

---

1 In computing, **impatience** is a virtue that leads to the development of faster algorithms.
2 Yes, this is an Indiana Jones reference.

1.2 Terminology

Here are some important definitions:

- A **numerical method** is any mathematical technique used to approximate a solution to a mathematical problem.

  Common examples of numerical methods you may already know include Newton's method for root-finding and Gaussian elimination for solving systems of linear equations.

- An **analytical solution** is a closed form expression for unknown variables in terms of the known variables.

  For example, suppose we want to solve the problem

  $$0 = ax^2 + bx + c$$

  for a given (known) a, b, and c. The quadratic formula tells us the solutions are

  $$x = \frac{-b \pm \sqrt{b^2 - 4ac}}{2a}.$$

  Each of these solutions is an analytical solution to the problem.

- A **numerical solution** is a number that approximates a solution to a mathematical problem in one particular instance.

For the example above for finding the roots of a quadratic polynomial, to use a numerical method such as Newton's method, we would need to start with a specified a, b, and c. Suppose we choose $a = 1$, $b = -2$, $c = -1$, and ran Newton's method with an initial guess of $x_0 = 0$. This returns the numerical solution of $x = -0.414213562373095$.

There are two clear disadvantages to numerical solutions compared to analytic solutions. First, they only work for a particular instance of a problem, and second, they are not as accurate. It turns out this solution is accurate to 16 digits (which is approximately the standard number of digits a computer stores for any number), but if you needed accuracy to 20 digits then you need to go through some serious work to get it. But many other numerical methods will only give 'a few' digits of accuracy in a reasonable amount of time.

On the other hand, there is a clear advantage to using numerical methods in that they can solve many problems that we cannot solve analytically. For example, can you analytically find the solution to

$$x^2 = e^x?$$

Probably you cannot. But if we look at the plots of $y = e^x$ and $y = x^2$ in Figure 1.1, it is clear that a solution exists. If we run Newton's method to find the zero of $x^2 - e^x$, it takes no time at all to arrive at the approximation (correct to 16 digits) $x = -0.703467422498392$. In this sense, numerical methods can be an enabling technology.

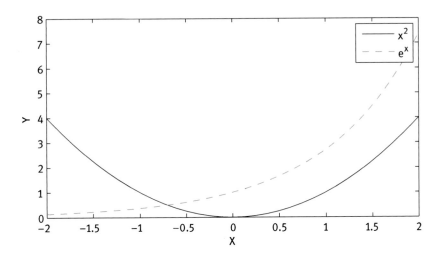

Fig. 1.1. Plots of e^x and x^2.

The plot in Figure 1.1 was created with the following commands:

```
>> x = linspace(-2,2,100);
>> y1 = x.^2;
>> y2 = exp(x);
>> plot(x,y1,'k-',x,y2,'r--')
>> xlabel('X')
>> ylabel('Y')
>> legend('x^2','e^x')
```

Some notes on these commands:
- The function `linspace(a,b,n)` creates a vector of n equally spaced points from a to b.
- In the definition of `y1`, we use a period in front of the power symbol. This denotes a 'vector operation'. Since x is a vector, this will do the operation component-wise, and so `y1` will be a vector of the squares of the components of x.
- `exp` is the exponential operator, so `exp(x)` = e^x. This operator does vector operations as well.
- The `plot` command plots x vs. y values (each given as a vector of values). The third argument is determines the style for plotting. You can plot more than one function into the same plot by listing additional pairs of vectors.
- The last three lines add axis labels and a legend.

1.3 Convergence terminology

For a given mathematical problem, assume there is a solution, and call it u. If we use a numerical algorithm to approximate u, then we will get a numerical solution, call it $\tilde{u}$ (it is extremely rare for $u = \tilde{u}$). The fundamental question is: How close is $\tilde{u}$ to u? It is our job, and the purpose of this course, to learn how to quantify this difference for various common numerical algorithms.

In many cases, the error in an approximation depends on a parameter. For example, in Newton's method, the error typically depends on how many iterations are performed. If one is approximating a derivative with $f'(x)$ by calculating $\frac{f(x+h)-f(x)}{h}$ for a fixed h, the error will naturally depend on h. Hence in this case, we will want to quantify the error in terms of h; that is, we want to be able to write

$$\left| \frac{f(x+h)-f(x)}{h} - f'(x) \right| \le Ch^k,$$

where C is a problem dependent constant (independent of h and k), and we wish k to be as large as possible. If $k > 0$, then as h decreases, we are assured the error will go to 0. The larger k is, the faster it will go to zero.

Definition 1 (Big O notation). *Suppose u is the true solution to a mathematical problem, and $\tilde{u}(h)$ is an approximation to the solution that depends on a parameter h. If it holds that*

$$|u - \tilde{u}(h)| \le Ch^k,$$

with C being a constant independent of h and k, then we write

$$|u - \tilde{u}(h)| = O(h^k).$$

This is interpreted as "The error is on the order of h^k."

For first order convergence ($k = 1$), the error is reduced proportional to the reduction of h. In other words, if h gets cut in half, you can expect the error to be approximately cut in half also. For second order convergence, however, if h gets cut in half, the error gets cut in fourth, which is obviously much better.

Example 2. *Suppose we have an algorithm where the error depends only on h, and for a sequence of h's $\{1, 1/2, 1/4, 1/8, 1/16, 1/32, \ldots\}$, the sequence of errors*

$$10, 5, 2.5, 1.25, 0.625, 0.3125$$

converges with first order accuracy, i.e. O(h). This is because when h gets cut in half, so do the errors. The sequence of errors

$$100, 25, 6.25, 1.5625, 0.390625, 0.09765625$$

converges with second order accuracy, i.e. $O(h^2)$, since the errors get cut by $4 = 2^2$, when h is cut in half.

Notice that even though for larger h the errors for linear convergence were smaller, as h gets smaller, the errors for the second-order convergence sequence become much better than the linear convergence errors.

In the example above, it was clear that the exponents of h were 1 and 2, but in general the k in $O(h^k)$ need not be an integer. To approximate k from two data points (h_1, e_1), (h_2, e_2), we treat the error bound as a close approximation, and start with

$$e \approx Ch^k$$

with C independent of h and k. Then we have that

$$e_1 \approx Ch_1^k, \quad e_2 \approx Ch_2^k.$$

Solving for C in both equations and setting them equal gives

$$\frac{e_1}{h_1^k} \approx \frac{e_2}{h_2^k} \implies \frac{h_2}{h_1}^k \approx \frac{e_2}{e_1} \implies k \log\left(\frac{h_2}{h_1}\right) \approx \log\left(\frac{e_2}{e_1}\right) \implies k \approx \frac{\log\left(\frac{e_2}{e_1}\right)}{\log\left(\frac{h_2}{h_1}\right)}.$$

Given a sequence of h and corresponding errors, we calculate k for each successive error, and typically k will converge to a number.

Remark 3. *Here we defined $O(h^k)$ for a parameter h going to zero to represent accuracy of a method. Later we will have algorithms where we would like to describe the complexity of an algorithm depending on an integer n, which typically stands for the number of elements in the input. Complexity can be the time to run the algorithm, or (equivalently) the total number of floating point operations the computer must make to complete the task. For this we use the notation $O(f(n))$ where f is a function (for us often a polynomial) in n that describes how the complexity of the algorithm grows with n.*

For example: finding the largest element in a list of n numbers requires you to look at each number once. This algorithm would have the complexity $O(n)$, and the time to run it will grow linearly with the number of elements in your list. If somebody had already sorted the list for you, you can just look at the first element. That algorithm would be $O(1)$ because it always takes the same amount of time (independent of n).

1.4 Exercises

1. Create a plot of $y = x$, $y = x^2$, and $y = \sin(x)$ on $[-1, 1]$, complete with axis labels and a legend.
2. Suppose error is a function of h, and h decreases via the sequence

$$\{1, 1/2, 1/4, 1/8, 1/16, 1/32, \ldots\}.$$

Classify the rate of convergence (i.e. the k in $O(h^k)$) for the corresponding sequence of errors:

a) $1, \dfrac{1}{2}, \dfrac{1}{4}, \dfrac{1}{8}, \dfrac{1}{16}, \dfrac{1}{32}, \ldots;$

b) $1, \dfrac{1}{8}, \dfrac{1}{64}, \dfrac{1}{256}, \dfrac{1}{2048}, \dfrac{1}{16384}, \ldots;$

c) $1, 0.3536, 0.1250, 0.0442, 0.01563, 0.005524, 0.001953, \ldots.$

2 Computer representation of numbers and roundoff error

In this chapter, we will introduce the notion and consequences of a finite number system. Each number in a computer must be physically stored, and therefore a computer can only hold a finite number of digits for any number. Decades of research and experimentation has led us to a (usually) reasonable approximation of numbers by representing them with (about) sixteen digits of accuracy. While this approximation may seem at first to be perfectly reasonable, we will explore its consequences in this chapter. Moreover, given that we must use a finite number system, we will investigate how to not make terrible mistakes.

2.1 Examples of the effects of roundoff error

To motivate the need to study computer representation of numbers, let's consider first some examples taken from MATLAB – but we note the same thing happens in C, Java, etc.:

1. The order in which you add numbers on a computer makes a difference!

    ```
    >> 1 + 1e-16 + 1e-16 + 1e-16 + 1e-16 + 1e-16 + 1e-16 + 1e-16 ...
       + 1e-16 + 1e-16 + 1e-16
    ans =
           1

    >> 1e-16 + 1e-16 + 1e-16 + 1e-16 + 1e-16 + 1e-16 + 1e-16 + 1e-16 ...
       + 1e-16 + 1e-16 + 1
    ans =
       1.000000000000001
    ```

    Note: AAAeBBB is a common notation for a floating point number with the value $AAA \times 10^{BBB}$. So 1e-16 = 10^{-16}.

    As we will see later in this chapter, the computer stores about 16 base 10 digits for each number; this means we get 15 digits after the first nonzero digit of a number. Hence if you try to add 1e-16 to 1, there is nowhere for the computer to store the 1e-16 since it is the 17th digit of a number starting with 1. It does not matter how many times you do it, it just gets lost each time, since operations are always done from left to right. So even if we add 1e-16 to 1, 10 times in a row, we get back exactly 1. However, if we add 10 1e-16's together first, then add the 1, these small numbers get a chance to combine to 1e-15, which is big enough not to be lost when added to 1.

2. Consider

$$f(x) = \frac{e^x - e^{-x}}{x}.$$

Suppose we wish to calculate

$$\lim_{x \to 0} f(x).$$

By L'Hopital's theorem, we can easily determine the answer to be 2. However, how might one do this on a computer? A limit is an infinite process, and moreover it requires some analysis to get an answer. Hence on a computer one is basically left with the option of choosing small x's and plugging them into f. The table below shows what we get back from MATLAB by doing so.

Table 2.1. Unstable limit computation.

x	$\dfrac{e^x - e^{-x}}{x}$
1e-6	1.999999999946489
1e-7	1.999999998947288
1e-8	1.999999987845058
1e-9	2.000000054458440
1e-10	2.000000165480742
1e-11	2.000000165480742
1e-12	2.000066778862220
1e-13	1.999511667349907
1e-14	1.998401444325282
1e-15	2.109423746787797
1e-16	1.110223024625157
1e-17	0

Moreover, if we choose x any smaller than 1e-17, we still get 0. The main numerical issue here is, as we will learn, subtracting two nearly equal numbers on a computer is bad and can lead to large errors.

Interestingly, if we create the limit table using a mathematically equivalent expression for $f(x)$, we can get a much better answer. Recall (that is, this is a fact which we hope you have seen before): the exponential function is defined by

$$e^x = \sum_{n=0}^{\infty} \frac{x^n}{n!} = 1 + x + \frac{x^2}{2!} + \frac{x^3}{3!} + \frac{x^4}{4!} + \frac{x^5}{5!} + \cdots$$

Using this definition, we calculate

$$\frac{e^x - e^{-x}}{x} = \frac{2x + 2\frac{x^3}{3!} + 2\frac{x^5}{5!} + 2\frac{x^7}{7!} + \cdots}{x} = 2 + 2\frac{x^2}{3!} + 2\frac{x^4}{5!} + 2\frac{x^6}{7!} + \cdots$$

Notice that if $|x| < 10^{-4}$, then $2\frac{x^4}{5!} < 10^{-20}$ and therefore this term (and all the ones after it in the sum) will not affect the calculation of the sum in any of the first

16 digits, which is all the computer stores. Hence we have that

$$\frac{e^x - e^{-x}}{x} \approx 2 + \frac{x^2}{3}$$

provided $|x| < 10^{-4}$, and we can expect this approximate of $f(x)$ to be accurate to 16 digits. Re-calculating a limit table based on this mathematically equivalent expression provides much better results, as can be seen in Table 2.2, which shows the limit is clearly 2.

Table 2.2. Stable limit computation.

x	$2 + \dfrac{x^2}{3}$
1e-6	2.000000000000334
1e-7	2.000000000000004
1e-8	2.000000000000000
1e-9	2.000000000000000
1e-10	2.000000000000000
1e-11	2.000000000000000
1e-12	2.000000000000000
1e-13	2.000000000000000
1e-14	2.000000000000000
1e-15	2.000000000000000
1e-16	2.000000000000000
1e-17	2.000000000000000

3. We learn in Calculus II that the integral

$$\int_{1}^{\infty} \frac{1}{x} = \infty.$$

Note that since this is a decreasing, concave up function, approximating it with the left rectangle rule gives an over-approximation of the integral. That is,

$$\sum_{n=1}^{\infty} \frac{1}{n} > \int_{1}^{\infty} \frac{1}{x},$$

and so we have that

$$\sum_{n=1}^{\infty} \frac{1}{n} = \infty.$$

But if we calculate this sum in MATLAB, we converge to a finite number instead of ∞. As mentioned above, the computer can only store (about) 16 digits. Since the running sum will be greater than 1, any number smaller than 1e-16 will be lost due to roundoff error. Thus, if we add in order, we get that

$$\sum_{n=1}^{\infty} \frac{1}{n} = \text{(in MATLAB)} \sum_{n=1}^{10^{16}} \frac{1}{n} < \infty.$$

Hence numerical error has caused a sum which should be infinite to be finite!

2.2 Binary numbers

Computers and software allow us to work in base 10, but behind the scenes everything is done in base 2. This is because numbers are stored in computer memory (essentially) as 'voltage on' (1) or 'voltage off' (0). Hence it is natural to represent numbers in their base 2, or binary, representation. To explain this, let us start with base 10 (or decimal) number system. In base 10, the number 12.625 can be expanded into powers of 10, each multiplied by a coefficient:

$$12.625 = 1 \times 10^1 + 2 \times 10^0 + 6 \times 10^{-1} + 2 \times 10^{-2} + 5 \times 10^{-3}.$$

It should be intuitive that the coefficients of the powers of 10 must be digits between 0 and 9, because if we have a coefficient of 10 or more, then we would have another power of 10 in the next coefficient to the left. Also, the decimal point goes between the coefficients of 10^0 and 10^{-1}.

Base 2 numbers work in an analogous fashion. First, note that it only makes sense to have digits of 0 and 1, for the same reason that digits in base 10 must be 0 through 9. Also, the decimal point goes between the coefficients of 2^0 and 2^{-1}. Hence in base 2 we have, for example, that

$$(11.001)_{\text{base2}} = 1 \times 2^1 + 1 \times 2^0 + 0 \times 2^{-1} + 0 \times 2^{-2} + 1 \times 2^{-3} = 2 + 1 + \frac{1}{8} = 3.125.$$

Converting a base 2 number to a base 10 number is nothing more than expanding it in powers of 2. To get an intuition for this, consider Table 2.3, which converts the base 10 numbers 1 through 10.

The following algorithm will convert a base 10 number to a base 2 number. Note this is not the most efficient computational algorithm, but perhaps it is the easiest to understand for beginners.

Given a base 10 decimal d,

1. find the biggest power p of 2 such that $2^p \le d$ but $2^{p+1} > d$, and save the index number p;
2. set $d = d - 2^p$;
3. if d is 0 or very small (compared to the d you started with) or the process has been repeated "enough", then stop. Otherwise repeat steps 1 and 2.

Table 2.3. Binary representation of the numbers from 1 to 10.

Base 10 representation	Base 2 representation
1	1
2	10
3	11
4	100
5	101
6	110
7	111
8	1000
9	1001
10	1010

Finally, write down your number as 1's and 0's, putting 1's in the entries where you save the index numbers, and 0 everywhere else.

Example 4. *Convert the base* 10 *number d* = 11.5625 *to base* 2.

Step 1: $8 = 2^3$ *is the biggest power of 2 less than d, so save 3, and set d* = 11.5625 – 8 = 3.5625.

Step 2: $2 = 2^1$ *is the biggest power of 2 less than d, so save 1, and set d* = 3.5625 – 2 = 1.5625.

Step 3: $1 = 2^0$ *is the biggest power of 2 less than d, so save 0, and set d* = 1.5625 – 1 = 0.5625.

Step 4: $\frac{1}{2} = 2^{-1}$ *is the biggest power of 2 less than d, so save* –1, *and set d* = 0.5625 – 0.5 = 0.0625.

Step 5: $\frac{1}{16} = 2^{-4}$ *is the biggest power of 2 less than d, so save* –4, *and set d* = 0.0625 – 0.0625 = 0.

Thus the process has terminated, since d = 0. *Our base* 2 *number thus has* 1*'s in the* 3, 1, 0, –1, *and* –4 *places, so we get*

$$(11.5625)_{\text{base10}} = (1011.1001)_{\text{base2}}.$$

Of course, not every base 10 number will terminate to a finite number of base 2 digits. For most computer number systems, we have a total of 53 base 2 digits to represent a base 10 number.

We now introduce the notion of **standard binary form**, which can be considered analogous to exponential formatting in base 10. The idea here is that every binary number, except 0, can be represented as

$$x = 1.b_1 b_2 b_3 \cdots \times 2^{\text{exponent}},$$

where each b_i is a 0 or a 1, and the exponent (a positive or negative decimal number) is adjusted so that the only number to the left of the decimal is a single 1. Some examples are given in the table below.

Table 2.4. Examples for the standard binary form.

Base 2	Standard binary form
1.101	1.101×2^0
1011.1101	1.0111101×2^3
0.000011101	1.1101×2^{-5}

2.3 64 bit floating point numbers

The most common, by far, computer number representation system is the 64-bit 'double' floating point number system. This is the default used by all major mathematical and computational software. In extreme cases, one can use 32 or 128 bit number systems, but that is a discussion for later (later, as in: not in this book), as first we must learn the basics. Each 'bit' on a computer is a 0 or a 1, and each number on a computer is represented by 64 0's and 1's. If we assume each number is in standard binary form, then the important information for each number is i) sign of the number, ii) exponent, iii) the digits after the decimal point. Note that the number 0 is an exception and is treated as a special case for the number system; we will not consider it here.

The IEEE standard divides up the 64 bits as follows:
- 1 bit sign: 0 for positive, 1 for negative;
- 11 bit exponent: the base 2 representation of (standard binary form exponent + 1023);
- 52 bit mantissa: the first 52 digits after decimal point from standard binary form.

The reason for the "shift" (sometimes also called bias) of 1023 in the exponent is so that the computer does not have to store a sign for the exponent (more numbers can be stored this way). The computer knows internally that the number is shifted, and knows how to handle it.

Example 5. *Convert the base 10 number d = 11.5625 to 64 bit double floating point representation.*

From a previous example, we know that $11.5625 = (1011.1001)_{base2}$, *and so has standard binary representation of* 1.0111001×2^3. *Hence we immediately know that*

sign bit = 0

mantissa = 0111001000000000000000000000000000000000000000000000

For the exponent, we need the binary representation of $(3 + 1023) = 1026 = 1024 + 2$, *and thus*

exponent = 10000000010

There are some obvious consequences to this number system:

1. There is a biggest and smallest positive representable number!
   - Since the exponent can hold 11 bits total (which means the base 10 numbers 0 to 2047), due to the shift the biggest positive unshifted number it can hold is 1024, and the biggest negative unshifted number it can hold is −1023. However, the unshifted numbers 1024 and −1023 have special meanings (e.g. representing 0 and infinity), and so the smallest and largest workable exponents are −1022 and 1023. This means the largest number that a computer can hold is
   $$n_{max} = 1.1111111 \cdots 1 \times 2^{1023} \approx 10^{308}$$
   and similarly the smallest positive representable number is
   $$n_{min} = 1.00000 \cdots 0 \times 2^{-1022} \approx 10^{-308}$$
   - Having these numbers as upper and lower bounds on positive representable numbers is generally not a problem. Usually, if one needs to deal with numbers larger or smaller than this, the entire problem can be rescaled into units that are representable.
2. The relative spacing between 2 floating point numbers is $2^{-52} \approx 2.22 \times 10^{-16}$.
   - This relative spacing between numbers is generally referred to as **machine epsilon**, and we will denote it by $\epsilon_{mach} = 2^{-52}$.
   - Given a number $d = 1.b_1 b_2 \cdots b_{52} \times 2^{exponent}$, the smallest increment we can add to it is in the 52nd digit of the mantissa, which is $2^{-52} \times 2^{exponent}$. Any number smaller would be after the 52nd digit in the mantissa.
   - As there is spacing between two floating point numbers, any real number between two floating point numbers must be rounded (to the nearest one). This means the maximum relative error in representing a number on the computer is about 1.11×10^{-16}. Thus, we can expect 16 digits of accuracy if we enter a number into the computer.
   - Although it is usually enough to have 16 digits of accuracy, in some situations this is not sufficient. Since we often rely on computer calculations to let us know a plane is going to fly or a boat is going to float or a reactor will not melt down, it is critical to know when computer arithmetic error can cause a problem and how to avoid it.

There are two main types of catastrophic computer arithmetic error: adding large and small numbers together, and subtracting two nearly equal numbers. We will describe each of these issues now.

2.3.1 Avoid adding large and small numbers

As we saw in example 1 in this chapter, if we add 1 to 10^{-16}, the 10^{-16} does not change the 1 at all. Additionally, the next computer representable number after 1 is $1 + 2^{-52} = 1 + 2.22 \times 10^{-16}$. Since $1 + 10^{-16}$ is closer to 1 than it is to $1 + 2.22 \times 10^{-16}$, it gets rounded to 1 - leaving the 10^{-16} to be lost forever.

We have seen this effect in the example at the beginning of this chapter when repeatedly adding 1e-16 to 1. Theoretically speaking, addition in floating point computation is not associative, meaning $(A + B) + C \neq A + (B + C)$ due to rounding.

One way to minimize this type of error when adding several numbers is to add from smallest to largest (if they all have the same sign), and to use factorizations that lessen the problem. There are other more complicated ways to deal with this kind of error that is out of the scope of this book, for example the "Kahan Summation Formula".

2.3.2 Subtracting two nearly equal numbers is bad

The issue here is that insignificant digits can become significant digits, and the problem is illustrated in example 2 in this chapter. Consider the following MATLAB command and output:

```
>> 1 + 1e-15 - 1
ans =
      1.110223024625157e-15
```

Clearly, the answer should be 10^{-15}, but we do not get that at all. It is true that the digits of accuracy in the subtraction operation is 16, but there is a potential problem with the "garbage" digits 110223024625157 (these digits arise from rounding error). If we are calculating a limit, for example, they could play a role:

Consider using the computer to find the derivative of $f(x) = x$ at $x = 1$. We all know the answer is $f'(1) = 1$, but suppose for a moment we do not know that and wish to calculate an approximation. The definition of the derivative tells us

$$f'(1) = \lim_{h \to 0} \frac{f(1 + h) - f(1)}{h}.$$

It might seem reasonable to just pick a very small h to get a good answer, but this is a bad idea! Consider the following MATLAB commands, which plug in values of h from 10^{-1} to 10^{-20}. When $h = 10^{-15}$, we see the "garbage" digits have become significant and affect the second significant digit! For smaller values of h, $1 + h$ gives back 1, and so the derivative is 0.

```
>> h = 10.^-[1:20]';
>> fp = ((1+h) - 1) ./ h;
>> disp([h, fp]);
   0.100000000000000   1.000000000000001
   0.010000000000000   1.000000000000001
   0.001000000000000   0.999999999999890
   0.000100000000000   0.999999999999890
   0.000010000000000   1.000000000006551
   0.000001000000000   0.999999999917733
   0.000000100000000   1.000000000583867
   0.000000010000000   0.999999993922529
   0.000000001000000   1.000000082740371
   0.000000000100000   1.000000082740371
   0.000000000010000   1.000000082740371
   0.000000000001000   1.000088900582341
   0.000000000000100   0.999200722162641
   0.000000000000010   0.999200722162641
   0.000000000000001   1.110223024625157
   0.000000000000000                   0
   0.000000000000000                   0
   0.000000000000000                   0
   0.000000000000000                   0
   0.000000000000000                   0
```

We will discuss in detail in the next chapter more accurate ways to approximate derivatives.

Finally, there is one other fact about floating point numbers to be aware of. What do you expect the following program to do?

```
i=1.0
while i~=0.0
  i=i-0.1
end
```

The operator "~=" means "not equal" in MATLAB. We would expect the loop to be counting down from 1.0 in steps of 0.1 until we reach 0, right? No, it turns out that we are running an endless loop counting downwards because i is never exactly equal to 0.

Rule: Never compare floating point numbers for equality ("==" or "~="). Instead, for example, use a comparison such as

```
abs(i)< 1e-12
```

to replace the comparison i~=0.0 from the above code.

2.4 Exercises

1. Convert the binary number 1101101.1011 to decimal format.
2. Convert the decimal number −66.125 to binary format. What is its 64-bit floating point representation?
3. Show that the decimal number 0.1 cannot be represented exactly as a finite binary number. Use this fact to explain that "0.1 * 3 == 0.3" returns 0 (meaning false) in MATLAB.
4. Write a MATLAB function tobinary(n) that, given a whole number, outputs its base 2 representation. The output should be a vector with 0s and 1s. Example:

```
>> tobinary(34)
ans =
    1   0   0   0   1   0
```

   Hint: For finding the power p of 2 in the algorithm it helps to know that you can just check the values p = pmax, pmax-1, ..., 0 where pmax is the logarithm in base 2 rounded down: pmax=floor(log2(n)). Then either add a zero or a one to the output vector for each p.
5. What is the minimum and maximum distance between two adjacent floating point numbers?
6. What is the best way to numerically evaluate $\|x\|_2 = \sqrt{\sum_{n=1}^{N} x_n^2}$? (i.e., is there a best order for the addition?)
7. Consider the function

$$g(x) = \frac{e^x - 1}{x}.$$

   (a) Find $\lim_{x \to 0} g(x)$ analytically.
   (b) Write a MATLAB program to calculate $g(x)$ for $x = 10^{-1}, 10^{-2}, ..., 10^{-15}$. Explain why the analytical limit and the numerical 'limit' do not agree.
   (c) Extend your calculation to $x = 10^{-16}, 10^{-17}, ..., 10^{-20}$, and explain this behavior.
8. The polynomial $(x - 2)^6$ is zero when $x = 2$ but positive everywhere else. Plot both this function, and its decomposition $x^6 - 12x^5 + 60x^4 - 160x^3 + 240x^2 - 192x + 64$ near $x = 2$, on [1.99,2.01] using 10,000 points. Explain the differences in the plots.
9. What is next biggest computer representable number after 1 (assuming 64 double floating point numbers)? What about after 4096? What about after $\frac{1}{8}$?

3 Solving linear systems of equations

The need to solve systems of linear equations arises across nearly all of engineering and science, business, statistics, economics, etc. In the chapters that follow, the need arises in numerical methods for solving boundary value differential equations, and as part of nonlinear equation solvers. We present in this chapter the method of Gaussian elimination to solve a system of linear equations with n equations and n unknowns. Gaussian elimination is an important first algorithm when learning about solving linear systems. For small, dense matrices (dense meaning mostly nonzero entries in the matrix), this algorithm and variants of it are typically the best that can be done. However, for large sparse linear systems that arise from finite element/difference discretizations of differential equations, iterative solvers such as Conjugate Gradient and GMRES are typically better choices. We do not discuss such algorithms herein, but we refer the interested reader to the literature, of which there is a vast amount, but in particular the books 'Numerical Linear Algebra' by Layton and Sussman (2014, Lulu publishing), and 'Numerical Linear Algebra' by Trefethen and Bau (SIAM, 1995).

3.1 Linear systems of equations and solvability

A linear equation of 1 variable takes the form

$$ax = b,$$

where a and b are given, and we want to determine the unknown value x. Solving this equation is easy, provided $a \neq 0$. If $a = 0$, however, there is no solution when $b \neq 0$, and infinitely many solutions if $b = 0$.

A linear system of 2 variables and 2 equations can be represented by

$$a_{11}x_1 + a_{12}x_2 = b_1$$
$$a_{21}x_1 + a_{22}x_2 = b_2$$

where the unknowns are x_1 and x_2, and the other 6 numbers are given. Most college students have solved equations like this in high-school, and most of the time a solution can be found by solving for one of the x's in terms of the other from one equation, plugging that into the other equation to determine one of the x's, then determining the remaining x using the known one. However, this procedure can fail: a solution need not exist, or there could also be infinitely many solutions, and in both of these cases this procedure will not produce a unique solution. Consider the following examples.

Example 6. *There are no solutions to the linear system*

$$2x_1 + x_2 = 0$$
$$2x_1 + x_2 = 1.$$

These equations are not consistent, and therefore no solution can exist.

Example 7. *There are infinitely many solutions to the linear system*

$$2x_1 + x_2 = 0$$
$$4x_1 + 2x_2 = 0.$$

Here, the equations can be reduced to the exact same equation. Thus, we have 2 unknowns, x_1 and x_2, but only one equation. Thus, any point on the line $2x_1 + x_2 = 0$ will solve the system.

A geometric interpretation of solving linear systems is that if you were to plot each equation, the solution is where they intersect. Hence, most of the time, two lines intersect at a point. However, the lines could be parallel and never intersect (this is the inconsistent case), or they could be the same line and intersect at every point on the line. This same idea can be extended to larger systems, and generally speaking, an n-equation, n-unknown consistent linear system will have a unique solution provided no equation is 'repeated'.

In the 2 × 2 case, it is easy to see when an equation is repeated or inconsistent, but for larger systems it is not always so easy. For example, consider the system of equations

$$2x_1 + x_2 = 0$$
$$4x_2 + 2x_3 = 1$$
$$2x_1 + 5x_2 + 2x_3 = 1.$$

Here, the third equation is merely the first two equations added together. Hence, it is not inconsistent, but it does not add anything to the system – if the first two equations are satisfied, the third will automatically be satisfied. Two consistent equations with 3 unknowns means infinitely many solutions (two planes intersecting along a line).

It will be more convenient to use matrix-vector notation to represent linear systems of equations. For example, the 3-equation, 3-unknown system above can be equivalently written as

$$\mathbf{Ax} = \begin{pmatrix} 2 & 1 & 0 \\ 0 & 4 & 2 \\ 2 & 5 & 2 \end{pmatrix} \begin{pmatrix} x_1 \\ x_2 \\ x_3 \end{pmatrix} = \begin{pmatrix} 0 \\ 1 \\ 1 \end{pmatrix} = \mathbf{b}.$$

If we are going to try to use a numerical algorithm to find a solution to a system of equations, it would be nice to know up front whether or not a unique solution exists for a particular linear system. It turns out that there are many ways to determine this for small systems (say, $n < 10{,}000$), for example if the determinant of the matrix of coefficients is nonzero, then a solution exists and is unique. For larger systems, however, this determination can be very difficult. Fortunately, most linear systems that arise in science and engineering from common approaches such as finite element and finite difference methods are uniquely solvable, and if not, often a mistake has been made in the derivation of the coefficients or there is a different approach to the problem that would lead to a solvable system. Thus, we will assume for the rest of this chapter that all the square linear systems have a unique solution, but we will keep in mind this potential problem when we analyze the algorithms. As the definition below states, this assumption can be made in several equivalent ways.

Definition 8. *A square linear system* $\mathbf{A}\mathbf{x} = \mathbf{b}$ *is uniquely solvable if the matrix* $\mathbf{A}$ *is nonsingular. Nonsingular is equivalent to each of:*
– $\mathbf{A}$ *is invertible;*
– $\det(\mathbf{A}) \neq 0$;
– *No eigenvalue of* $\mathbf{A}$ *is zero;*
– *No singular value of* $\mathbf{A}$ *is zero;*
– *The rows of* $\mathbf{A}$ *are linearly independent;*
– *The columns of* $\mathbf{A}$ *are linearly independent.*

3.2 Solving triangular systems

A first step along the way to solving general linear systems of equations is solving 'triangular systems', which get their name because the matrix representation of their coefficients has only 0's either above or below the main diagonal (which goes from the top left to the bottom right). For such systems, finding solutions is straightforward, as we show in the following example.

Example 9 (Back-substitution for an upper triangular system). *We want to use back-substitution to solve the upper triangular system of equations*

$$\begin{pmatrix} 1 & 2 & 3 \\ 0 & 4 & 5 \\ 0 & 0 & 6 \end{pmatrix}\begin{pmatrix} x_1 \\ x_2 \\ x_3 \end{pmatrix} = \begin{pmatrix} 1 \\ -5 \\ -6 \end{pmatrix} \iff \begin{array}{r} x_1 + 2x_2 + 3x_3 = 1 \\ 4x_2 + 5x_3 = -5 \\ 6x_3 = -6. \end{array}$$

The last equation can be solved directly for x_3:

$$6x_3 = -6 \implies x_3 = -1.$$

Plug in x_3 *into the first two equations, and rewrite as a system of two equations:*

$$\begin{pmatrix} 1 & 2 \\ 0 & 4 \end{pmatrix}\begin{pmatrix} x_1 \\ x_2 \end{pmatrix} = \begin{pmatrix} 1 - 3x_3 \\ -5 - 5x_3 \end{pmatrix} = \begin{pmatrix} 1 - 3(-1) \\ -5 - 5(-1) \end{pmatrix} = \begin{pmatrix} 4 \\ 0 \end{pmatrix} \iff \begin{array}{r} x_1 + 2x_2 = 4 \\ 4x_2 = 0. \end{array}$$

Notice the new, smaller 2×2 system remains upper triangular. That is a key point to the algorithm, as it can now be repeated over and over until the matrix reduces to a scalar. Solving the second equation directly for x_2 gives $x_2 = 0$. Finally, plugging x_2 into the first equation and solving yields $x_1 = 4$.

This procedure can easily be made into a general algorithm: For an $n \times n$ upper triangular linear system, starting at the bottom diagonal, solve for the unknown x_n in that row, then plug it in the above rows. This leaves a $(n - 1) \times (n - 1)$ triangular linear system, and so you repeat the process until the solution is determined. A code to do this is as follows:

```
function [x] = BackSubstitution(A,b)
% solve the upper triangular system Ax=b using back-substitution

n= length(b);
x = zeros(n,1);

for j=n:-1:1
    % Check to see if the diagonal entry is zero
    if abs(A(j,j)) < 1e-15
        error('A is singular (diagonal entries of zero)')
    end

    % Compute solution component
    x(j) = b(j) / A(j,j);

    % Update the RHS vector
    for i=1:j-1
        b(i) = b(i) - A(i,j)*x(j);
    end
end
```

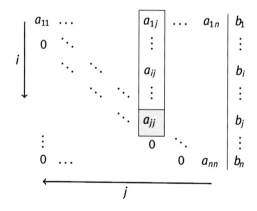

Fig. 3.1. Sketch of back-substitution. Starting from the bottom, in step j we solve for x_j using the diagonal element a_{jj} and then move the contributions of the column above into the vector b.

It can happen that this algorithm fails. If a diagonal entry is zero, then dividing by the diagonal entry is not possible. In such cases, it turns out that the rows of the matrix must have repetitions, which means there is not a unique solution to the problem - there will either be 0 or infinitely many. Note that the code above checks for this.

We can show that the computational work involved in the solving procedure is n^2 floating point operations, or flops; this is just the number of +, −, *, /. On step 1, we perform 1 division, and then $n - 1$ multiplications and $n - 1$ subtractions, for a total of $2n - 1$ flops. On step 2, we do the exact same thing, except on a matrix of size 1 smaller, so this gives $2n - 3$ flops. Repeating this gives:

$$\text{Work} = \text{Total flops} = (2n - 1) + (2n - 3) + \cdots + 1 = n^2$$

Given a lower triangular matrix $\mathbf{A}$ and a vector $\mathbf{b}$, how do we find a vector $\mathbf{x}$ such that $\mathbf{Ax} = \mathbf{b}$?

Example 10 (Lower triangular matrix). *A 4×4 lower triangular matrix:*

$$\mathbf{A} = \begin{pmatrix} a_{11} & 0 & 0 & 0 \\ a_{21} & a_{22} & 0 & 0 \\ a_{31} & a_{32} & a_{33} & 0 \\ a_{41} & a_{42} & a_{43} & a_{44} \end{pmatrix}.$$

The first equation depends only on x_1, thus if $a_{11} \neq 0$ then $x_1 = b_1/a_{11}$. We can then rewrite the system in terms of $x_2, x_3, \ldots, x_n$.

This process is the reverse of back-substitution, and is known as *forward-substitution*.

3.3 Gaussian elimination

We now move on to solving general square linear systems. Gaussian elimination is a 2 step algorithm to solve the system, the first of which is a manipulation of the matrix into an upper triangular matrix, and the second is to use back substitution to solve the linear system. To 'manipulate the matrix', we use the following row operations:
- Multiply a row by a constant (not equal to zero!).
- Replacing a row by a linear combination of that row with any other row. Since each row is an equation, if we add 2 rows together, the new equation must still hold.
- Swap any two rows. A matrix is just a set of linear equations, so their order is of no consequence to the solution.

The fact that none of these operations changes the solutions is known as Gauss' theorem.

Gaussian elimination is the process of manipulating a linear system to be upper triangular, using only the above row operations, and then solving the upper triangular system using back substitution. The way to do this is, column by column from left to right, add rows together in order to zero out entries below the diagonal. For each row, we always add to it the rescaled row holding the diagonal entry. When working on a particular column, the diagonal entry is called the pivot.

Note that it is possible for a zero to be the pivot, but if this is true, then there is no way to use it to zero out entries below it. In this case, we swap the row containing the pivot for a row below it that would not give a zero pivot. If there is no nonzero to swap with, then Gaussian elimination fails. However, in this case, it is known that the matrix A must have been singular, and a unique solution does not exist. If this happens and one does hope for a unique solution, then one must check carefully how the equations were created and input.

Let us proceed with an example:

Example 11 (Gaussian elimination). *Use Gaussian elimination to solve the system of equations $Ax = b$:*

$$\begin{pmatrix} 0 & 1 & 4 \\ 2 & 4 & 6 \\ 5 & 6 & 0 \end{pmatrix} \begin{pmatrix} x_1 \\ x_2 \\ x_3 \end{pmatrix} = \begin{pmatrix} 9 \\ 16 \\ 6 \end{pmatrix}.$$

To help with notation, let us first write this in augmented form (just put the right-hand side b next to the matrix):

$$\begin{array}{ccc|c} 0 & 1 & 4 & 9 \\ 2 & 4 & 6 & 16 \\ 5 & 6 & 0 & 6 \end{array}.$$

We begin in column one. Since a_{11} is zero, we need to swap one of the other rows with the first row. We pick the first one that has a non-zero entry:

$$\begin{array}{ccc|c} 0 & 1 & 4 & 9 \\ 2 & 4 & 6 & 16 \\ 5 & 6 & 0 & 6 \end{array} \quad \xrightarrow{\text{swap R1 and R2}} \quad \begin{array}{ccc|c} 2 & 4 & 6 & 16 \\ 0 & 1 & 4 & 9 \\ 5 & 6 & 0 & 6 \end{array}.$$

Now we need to eliminate all non-zero entries below the diagonal, which is in this case only row 3. We do this by adding row 1 multiplied by the factor $-\frac{5}{2}$ to row 3 to get:

$$\begin{array}{ccc|c} 2 & 4 & 6 & 16 \\ 0 & 1 & 4 & 9 \\ 5 & 6 & 0 & 6 \end{array} \quad \xrightarrow{R3 = R3 - \frac{5}{2} R1} \quad \begin{array}{ccc|c} 2 & 4 & 6 & 16 \\ 0 & 1 & 4 & 9 \\ 0 & -4 & -15 & -34 \end{array}.$$

With the first column completed, we need to eliminate the -4 in column 2 (the only entry below the diagonal that is non-zero). We do this by adding 4 times row 2 to row 3 (note that we must not use row 1, because this would reintroduce a value in the entry a_{31} that *we just eliminated):*

$$
\begin{array}{ccc|c}
2 & 4 & 6 & 16 \\
0 & 1 & 4 & 9 \\
0 & -4 & -15 & -34
\end{array}
\xrightarrow{R3\,=\,R3\,+\,4\,R2}
\begin{array}{ccc|c}
2 & 4 & 6 & 16 \\
0 & 1 & 4 & 9 \\
0 & 0 & 1 & 2
\end{array}.
$$

There is nothing to do in the third column, so step 1 of Gaussian Elimination is now completed. Back substitution will give us the solution $x_3 = 2, x_2 = 1, x_1 = 0$, so the solution to the linear system is:

$$
\mathbf{x} = \begin{pmatrix} 0 \\ 1 \\ 2 \end{pmatrix}.
$$

We now consider the algorithm and code for Gaussian elimination. We consider it here without pivoting, and leave adding this important step as an exercise.

Algorithm 12 (Gaussian elimination (without row swapping)).

for $k = 1$ to $n - 1$
    if $a_{kk} = 0$ then stop
    for $i = k + 1$ to n
        $m = a_{ik}/a_{kk}$
        for $j = k + 1$ to n
            $a_{ij} = a_{ij} - m \cdot a_{kj}$
        end
        $b_i = b_i - m \cdot b_k$
    end
end

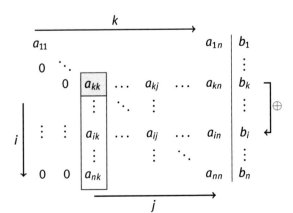

Fig. 3.2. Sketch of Gaussian elimination. In step k the element a_{kk} is used to eliminate the entries below. The entry a_{ik} becomes 0 after adding a multiple of the row k to the row i.

```
function [x] = GaussianElimination(A,b)

n = length(b);

for k=1:n-1
    % Check to see if the pivot is zero
    if abs(A(k,k)) < 1e-15;
        error('A has diagonal entries of zero')
    end

    % Apply transformation to remaining submatrix and RHS vector
    for i=k+1:n
        m = A(i,k)/A(k,k); %multiplier for current row i
        for j=k+1:n
            A(i,j) = A(i,j) - m*A(k,j);
        end
        b(i) = b(i) - m*b(k);
    end
end

% A is now upper triangular.  Use back substitution to solve
% the transformed problem.
x = BackSubstitution(A,b);
```

From the algorithm, we can estimate the work being done in terms of flops. On the kth time through the outer loop, we perform:
1. $n - k$ divisions;
2. $(n - k)(n - k)$ multiplications;
3. $(n - k)(n - k)$ subtractions.

Total operation count is $(n - k)[2(n - k) + 1]$ flops. The outer loop is performed for $k = 1$ to $(n - 1)$, so the total operation count is

$$\sum_{k=1}^{n-1}(n - k)[2(n - k) + 1] = 2\sum_{j=1}^{n-1} j(2j + 1) = 2\sum_{j=1}^{n-1} j^2 + \sum_{j=1}^{n-1} j.$$

Recall from Calculus that

$$\sum_{j=1}^{n} j^2 = \frac{n(n + 1)(2n + 1)}{6} \quad \text{and} \quad \sum_{j=1}^{n} j = \frac{n(n + 1)}{2}.$$

Using these two formulas we then see that the total operation count of Gaussian elimination is

$$2\sum_{j=1}^{n-1} j^2 + \sum_{j=1}^{n-1} j = 2\left[\frac{(n - 1)(n)(2n - 1)}{6}\right] + \frac{(n - 1)(n)}{2}.$$

Thus we see that Gaussian elimination requires approximately $2n^3/3$ flops, i.e. $O(n^3)$ complexity. This means that if you double the size of the matrix (e.g. go from 100×100 to 200×200), the time it takes to complete the algorithm will increase by a factor of 8.

3.4 The backslash operator

The system $\mathbf{Ax} = \mathbf{b}$ can be solved using Gaussian elimination via the command: x = A\b. For example:

```
>> A = [4 -2 -3; -6 7 1;8 3 4]

A =

        4      -2      -3
       -6       7       1
        8       3       4

>> b = [1;-1;2]

b =

        1
       -1
        2

>> x = A\b

x =

       0.2481
       0.0775
      -0.0543
```

The 'backslash' operator is very smart. In the worst case, it performs Gaussian elimination. However, some matrices are 'sparse', i.e. they are mostly filled with 0's. In this case, backslash will speed up dramatically because it will know it does not have to waste time operating with the zeros.

3.5 LU decomposition

Most matrices $\mathbf{A}$ can be decomposed into a product $\mathbf{A} = \mathbf{LU}$ where $\mathbf{L}$ is a lower triangular matrix and $\mathbf{U}$ is an upper triangular matrix. This is known as the LU decomposition (factorization) of $\mathbf{A}$. It turns out that such $\mathbf{L}$ and $\mathbf{U}$ can be found as the result of the Gaussian elimination algorithm (provided no pivoting is done, otherwise, the factorization will be $\mathbf{A} = \mathbf{PLU}$). The $\mathbf{U}$ is the upper triangular matrix you end up with from Gaussian elimination, and the $\mathbf{L}$ is a lower triangular matrix with 1's on the diagonal, and lower diagonal entries are equal to the negative the multiplier of the pivot used to zero out that entry in $\mathbf{A}$.

Example 13. *Create an LU factorization of* $\mathbf{A} = \left(\begin{smallmatrix} 1 & 2 \\ 3 & 4 \end{smallmatrix}\right)$.

We begin by applying the Gauss elimination procedure, which means to replace row 2 by: R2 = R2 − 3R1. This creates the upper triangular matrix

$$\mathbf{U} = \left(\begin{array}{cc} 1 & 2 \\ 0 & -2 \end{array}\right).$$

To create the lower triangular matrix L, we make a matrix with 1's on the diagonal, 0's above the diagonal, and for the 2,1 entry, we input negative the multiplier used to kill the 2,1 entry in the Gauss elimination procedure. Since the multiplier was −3, the 2,1 entry of L will be 3:

$$\mathbf{L} = \left(\begin{array}{cc} 1 & 0 \\ 3 & 1 \end{array}\right).$$

It is easy to check with matrix multiplication that $\mathbf{A} = \mathbf{LU}$.

The reason why LU factorization is important is that it dramatically reduces the solve time if the $\mathbf{L}$ and $\mathbf{U}$ are known ahead of time. Recall that solving $\mathbf{Ax} = \mathbf{b}$ using Gaussian elimination is $O(n^3)$ work, but solving with matrices $\mathbf{L}$ and $\mathbf{U}$ requires $O(n^2)$ work. Hence if we know $\mathbf{L}$ and $\mathbf{U}$ up front, we can do the following procedure to dramatically speed up solve time of $\mathbf{Ax} = \mathbf{b}$:

1. Solve $\mathbf{Ly} = \mathbf{b}$ for $\mathbf{y}$ using forward-substitution
2. Solve $\mathbf{Ux} = \mathbf{y}$ for $\mathbf{x}$ using back-substitution

This takes $2n^2$ work, which is much less than $O(n^3)$ work when n is large.

The LU decomposition of $\mathbf{A}$ is particularly useful if one has to do many solves with the same matrix $\mathbf{A}$ and different RHS vectors $\mathbf{b}$. This occurs, for example, in common discretizations of many partial differential equations such as the heat equation,

$$u_t = c\Delta u.$$

For the first solve, you must do the LU decomposition step which is $O(n^3)$, but every solve after is cheap. The syntax for LU decomposition in MATLAB is `[L,U]=lu(A)`.

3.6 Exercises

1. By hand, use Gaussian elimination to solve the system of equations

$$\mathbf{Ax} = \left(\begin{array}{ccc} 1 & 1 & 2 \\ 4 & 3 & 0 \\ 0 & 2 & 2 \end{array}\right)\left(\begin{array}{c} x_1 \\ x_2 \\ x_3 \end{array}\right) = \left(\begin{array}{c} 2 \\ 5 \\ 6 \end{array}\right) = \mathbf{b}.$$

Also, find an LU factorization of $\mathbf{A}$.

2. Adapt the Gaussian Elimination code from this chapter to use pivoting. Row swaps should be done if the pivot is smaller than 10^{-10} in absolute value. Use this code to solve the linear system

$$\begin{pmatrix} 0 & 1 & 4 \\ 2 & 4 & 6 \\ 5 & 6 & 0 \end{pmatrix} \begin{pmatrix} x_1 \\ x_2 \\ x_3 \end{pmatrix} = \begin{pmatrix} 8 \\ 15 \\ 5 \end{pmatrix}.$$

   Also solve this system using the backslash operator. Your code should give the same answer as backslash.

3. Consider the matrix that arises from solving boundary value ODEs from Chapter 4's 'convDiffReact' function, but without reaction or convection. This matrix A has -2 for each diagonal entry, 1's on each of the first upper and lower diagonals, and the rest of the matrix is all 0's. Such a matrix is considered to be sparse because is consists of mostly 0's.

   Let **b** be a vector of all ones. For n = 10, 100, 1000, 2000, 4000, 8000:

   (a) Create the matrix **A** by first creating a matrix of all 0's, then putting in the -2's on the main diagonal and 1's on the first upper and lower diagonals. Solve **Ax** = **b** using backslash, and time it using `tic` and `toc`. Hint: Using the `diag` command can help here.

   (b) Repeat the process, but using the sparse matrix format for **A**. You can do this inefficiently using the `sparse` command to turn your matrices from part a) into sparse format. A much better way to do this is to create the matrix using the `spdiags` command. The syntax to do this can be found in the help documentation of `spdiags` (type "help spdiags" on the command line). Is there a timing difference between solving sparse and full matrices using backslash? How much?

4. Consider the following example from Chemistry, where we mix toluene C_7H_8 with nitric acid HNO_3 to produce trinitrotoluene $C_7H_5O_6N_3$ (TNT). We want to find x, y, z, w, the proportions of each substance for a balanced reaction. That is, we need to balance the quantities of each element in the reaction

$$x\,C_7H_8 + y\,HNO_3 \rightarrow z\,C_7H_5O_6N_3 + w\,H_2O.$$

   The number of atoms present before the reaction is the same as the number after the reaction, and balancing each element gives a linear equation, e.g. for C we get that $7x = 7z$.

   Assuming we take $y = 100$, how much TNT and water will be produced by a balanced reaction?

4 Finite difference methods

Taylor series and Taylor's theorem play a fundamental role in finite difference method approximations of derivatives. We begin this chapter with a review of these important results.

Theorem 14 (Taylor series). *If a function f is infinitely differentiable, then for any specified expansion point x_0, it holds that*

$$f(x) = f(x_0) + f'(x_0)(x - x_0) + \frac{f''(x_0)}{2!}(x - x_0)^2 + \frac{f'''(x_0)}{3!}(x - x_0)^3 + \cdots$$

for every single real number x.

Theorem 15 (Taylor's theorem). *If a function f is $n+1$ times differentiable, then for any specified expansion point x_0, for every real number x there is a number c in the interval $[x, x_0]$ such that*

$$f(x) = f(x_0) + f'(x_0)(x-x_0) + \frac{f''(x_0)}{2!}(x-x_0)^2 + \cdots + \frac{f^{(n)}(x_0)}{n!}(x-x_0)^n + \frac{f^{(n+1)}(c)}{(n + 1)!}(x-x_0)^{n+1}.$$

The importance of Taylor's theorem is that we can pick a point x_0, and then represent the function $f(x)$, no matter how 'messy' it is, by the polynomial $f(x_0) + f'(x_0)(x-x_0) + \frac{f''(x_0)}{2!}(x - x_0)^2 + \cdots + \frac{f^{(n)}(x_0)}{n!}(x - x_0)^n$, and we know the error is given by the single term $\frac{f^{(n+1)}(c)}{(n+1)!}(x - x_0)^{n+1}$. Although we do not know c, we do know it lives in $[x, x_0]$. In many cases, this term can be estimated to give an upper bound on the error in approximating a function by a Taylor polynomial.

Example 16. *Use Taylor's theorem with $n = 2$ to approximate $\sin(x)$ at $x = 1.1$, using the expansion point $x_0 = 1$. Find a bound on the error using Taylor's theorem and compare the bound to the actual error.*
First, we calculate:

$$f(1) = \sin(1), \qquad f'(1) = \cos(1), \qquad f''(1) = -\sin(1).$$

Thus we can approximate f by: $f(x) \approx \sin(1) + \cos(1)(x - 1) - \frac{\sin(1)}{2}(x - 1)^2$ with error term $-\frac{\cos(c)}{6}(x - 1)^3$, where c is not known precisely, but is known to live in the interval $[1, 1.1]$. Hence at $x = 1.1$, we get the approximation

$$f(1.1) \approx \sin(1) + (0.1)\cos(1) - (0.01)\frac{\sin(1)}{2} = 0.891293860470671.$$

The error term was given by $-\frac{\cos(c)}{6}(x - 1)^3$, with c unknown but in the interval $[1, 1.1]$. We know that $|\cos(x)| \leq 1$, so we can upper bound the error by

$$\left| -\frac{\cos(c)}{6}(1.1 - 1)^3 \right| \leq \frac{1}{6}(0.001) = 1.6667e - 4.$$

Using MATLAB, we calculate this approximation, the actual answer, and the actual error:

```
>> approx=sin(1)+(0.1)*cos(1)-0.01/2*sin(1);
>> exact=sin(1.1);
>> error=approx-exact;
>> disp([approx, exact, error]);
   0.891293860470671   0.891207360061435   0.000086500409236
```

Note the actual error was smaller than the estimated error, but the estimated error was a guarantee of what the maximum error can be. Note that $\cos(1.1) \approx 0.45 \leq \cos(c) \leq 0.55 \approx \cos(1)$, so we can improve the error bound by using $|\cos(c)| \leq 0.55$, which gives the improved upper bound on the error estimate:

$$\left| -\frac{\cos(c)}{6}(1.1-1)^3 \right| \leq \frac{.55}{6}(0.001) = 9.1667e-5.$$

This is now quite close to the actual error.

4.1 Approximating the first derivative

4.1.1 Forward and backward differences

Consider a discretization of an interval [a,b] with N+1 equally spaced points, call them $x_0, x_1, \ldots, x_N$. Call the point spacing h, so that $h = x_{i+1} - x_i$. Suppose we are given function values

$$f(x_0), f(x_1), \ldots, f(x_N),$$

so we know just the points, but not the entire curve, as in the plot in Figure 4.1.

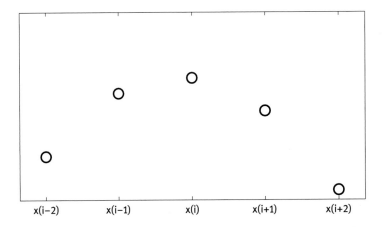

Fig. 4.1. Set of example points.

Suppose we want to know $f'(x_i)$ from just this limited information. Recalling the definition of the derivative is

$$f'(x) = \lim_{h \to 0} \frac{f(x + h) - f(x)}{h},$$

a first guess at approximating $f'(x_i)$ is to use one of

2 point forward difference: $\qquad f'(x_i) \approx \dfrac{f(x_{i+1}) - f(x_i)}{h},$

2 point backward difference: $\qquad f'(x_i) \approx \dfrac{f(x_i) - f(x_{i-1})}{h}.$

Illustrations of these ideas are shown in the figure below. These two finite difference approximations to the derivative are simply the slopes between the adjacent points.

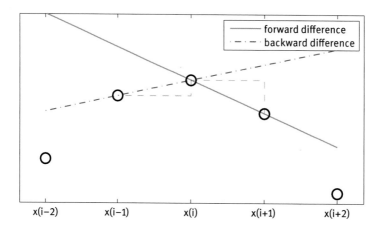

Fig. 4.2. The forward and backward finite difference approximations to $f'(x_i)$.

Important questions to ask about these approximations are "How accurate are they?" and "Is one of these approximations any better than the other?" Answers to these questions can be determined from Taylor series. Consider the spacing of the x points, h, to be a parameter. We expect that as h gets small, the forward and backward difference approximations should approach the true value of $f'(x_i)$ (assuming no significant roundoff error, which should not be a problem if h is not too small). Using Taylor's theorem with expansion point x_i and choosing $x = x_{i+1}$ gives us

$$f(x_{i+1}) = f(x_i) + f'(x_i)(x_{i+1} - x_i) + \frac{f''(c)}{2}(x_{i+1} - x_i)^2.$$

for some $x_i \le c \le x_{i+1}$. Then since $h = x_{i+1} - x_i$, this reduces to

$$f(x_{i+1}) = f(x_i) + hf'(x_i) + \frac{h^2 f''(c)}{2}.$$

Now some simple algebra yields

$$\frac{f(x_{i+1}) - f(x_i)}{h} - f'(x_i) = \frac{hf''(c)}{2}.$$

Note that the left hand side is precisely the difference between the forward difference approximation and the actual derivative $f'(x_i)$. Thus we have the following result:

Theorem 17. *Let f be a twice differentiable function in an interval containing x_i. Then the error in approximating $f'(x_i)$ with the forward difference approximation is bounded by*

$$\left| \frac{f(x_{i+1}) - f(x_i)}{h} - f'(x_i) \right| \leq Ch = O(h),$$

where C is a constant independent of h, and depends on the value of the second derivative of f near x_i.

The theorem above tells us the relationship between the error and the point spacing h is linear. In other words, if we cut h in half, we can expect the error to get cut (approximately) in half also.

For the backward difference approximation, we can do similar analysis as for forward difference, and will get the following result.

Theorem 18. *Let f be a twice differentiable function in an interval containing x_i. Then the error in approximating $f'(x_i)$ with the backward difference approximation is bounded by*

$$\left| \frac{f(x_i) - f(x_{i-1})}{h} - f'(x_i) \right| \leq Ch = O(h),$$

where C is a constant independent of h, and depends on the value of the second derivative of f near x_i.

Example 19. *Use the forward difference method, with varying h, to approximate the derivative of $f(x) = e^{\sin(x)}$ at $x = 0$. Furthermore, verify numerically that the error is $O(h)$.*

We type the following commands into MATLAB and get the following output for the forward difference calculations and the associated error (note $f'(0) = 1$)

```
>> h = (1/2).^[1:10];
>> fprime_fd = ( exp(sin(0+h))-exp(sin(0))) ./h;
>> error = fprime_fd - 1;
>> [h',fprime_fd',error',[0,error(1:end-1)./error(2:end)]']
```

```
ans =

   0.500000000000000   1.230292592884167   0.230292592884167                   0
   0.250000000000000   1.122785429776699   0.122785429776699   1.875569383948763
   0.125000000000000   1.062239501132755   0.062239501132755   1.972789427003951
   0.062500000000000   1.031218461877980   0.031218461877980   1.993676093845449
   0.031250000000000   1.015621121610778   0.015621121610778   1.998477616129762
   0.015625000000000   1.007812019185479   0.007812019185479   1.999626631718166
```

0.007812500000000	1.003906190146893	0.003906190146893	1.999907554857203
0.003906250000000	1.001953117533901	0.001953117533901	1.999977000406336
0.001953125000000	1.000976561567654	0.000976561567654	1.999994264153410
0.000976562500000	1.000488281133585	0.000488281133585	1.999998567392708

The first column is h, the second is the forward difference approximation, the third column is the error, and the fourth column is the ratios of the successive errors. To verify the method is O(h), we expect that the errors get cut in half when h gets cut in half, and the fourth column verifies this is true.

4.1.2 Centered difference

There are (many) more ways to approximate $f'(x_i)$ using finite differences. One thing to notice about the forward and backward differences is that if f has curvature, then for smaller h it must be true that one of the methods is an overestimate and the other is an underestimate. Then it makes sense that an average of the two methods might give a better approximation. After averaging, we get a formula that could also arise from using a finite difference of the values at x_{i-1} and x_{i+1}:

$$\textbf{2 point Centered difference:} \quad f'(x_i) \approx \frac{f(x_{i+1}) - f(x_{i-1})}{2h}.$$

A graphical illustration is given in Figure 4.3.

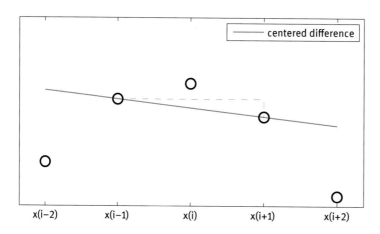

Fig. 4.3. The centered difference approximation to $f'(x_i)$.

Let us now compare the accuracy of the forward, backward, and centered difference approximations in an example.

Example 20. *Using h = 1 and h = 0.1, approximate $f'(1)$ with $f(x) = x^3$ using forward, backward, and centered difference methods. Note that the true solution is $f'(1) = 3$.*

For h = 1, we get the approximations

$$FD = \frac{f(2) - f(1)}{2 - 1} = 7 \qquad (|error| = 4) \tag{4.1}$$

$$BD = \frac{f(1) - f(0)}{1 - 0} = 1 \qquad (|error| = 2) \tag{4.2}$$

$$CD = \frac{f(2) - f(0)}{2 - 0} = 4 \qquad (|error| = 1). \tag{4.3}$$

For h = 0.1, we get the approximations

$$FD = \frac{f(1.1) - f(1)}{0.1} = 3.31 \qquad (|error| = 0.31) \tag{4.4}$$

$$BD = \frac{f(1) - f(0.9)}{0.1} = 2.71 \qquad (|error| = 0.29) \tag{4.5}$$

$$CD = \frac{f(1.1) - f(0.9)}{0.2} = 3.01 \qquad (|error| = 0.01). \tag{4.6}$$

We see in both examples that the centered difference is more accurate, and for the smaller h, centered difference is more than an order of magnitude more accurate.

The example above suggests that the centered difference method is significantly more accurate than the forward and backward difference methods. As it is dangerous to draw conclusions based on one example, let's look at what some mathematical analysis tells us:

Consider the Taylor series of a function f expanded about x_i:

$$f(x) = f(x_i) + f'(x_i)(x - x_i) + \frac{f''(x_i)}{2!}(x - x_i)^2 + \frac{f'''(x_i)}{3!}(x - x_i)^3 + \cdots .$$

Choose $x = x_{i+1}$ and then $x = x_{i-1}$ to get the two equations

$$f(x_{i+1}) = f(x_i) + hf'(x_i) + h^2\frac{f''(x_i)}{2!} + h^3\frac{f'''(x_i)}{3!} + \cdots$$

$$f(x_{i-1}) = f(x_i) - hf'(x_i) + h^2\frac{f''(x_i)}{2!} - h^3\frac{f'''(x_i)}{3!} + \cdots .$$

Subtracting the equations cancels out the odd terms on the right hand side, and gives us

$$f(x_{i+1}) - f(x_{i-1}) = 2hf'(x_i) + 2h^3\frac{f'''(x_i)}{3!} + \cdots$$

and thus using some algebra we get

$$\frac{f(x_{i+1}) - f(x_{i-1})}{2h} - f'(x_i) = h^2\frac{f'''(x_i)}{3!} + \cdots .$$

We now see that the leading error term in the centered difference approximation is $h^2\frac{f'''(x_i)}{3!}$. Thus, we repeat this procedure, but use Taylor's theorem to truncate at the third derivative terms. This gives us

$$\frac{f(x_{i+1}) - f(x_{i-1})}{2h} - f'(x_i) = h^2 \frac{f'''(c_1) + f'''(c_2)}{12}.$$

We have proved the following theorem:

Theorem 21. *Let f be a three times differentiable function in an interval containing x_i. Then the error in approximating $f'(x_i)$ with the centered difference approximation is bounded by*

$$\left| \frac{f(x_{i+1}) - f(x_{i-1})}{2h} - f'(x_i) \right| \leq Ch^2 = O(h^2),$$

where C can be considered a constant with respect to h, and it depends on the value of the third derivative of f near x_i.

This is a big deal! We have proven that the error in centered difference approximations is $O(h^2)$, whereas the error in forward and backward differences is $O(h)$! So if h gets cut by a factor of 10, we expect FD and BD errors to each get cut by a factor of 10, but the CD error will get cut by a factor of 100. Hence we can expect much better answers with CD, as we suspected.

Example 22. *Use the centered difference method, with varying h, to approximate the derivative of $f(x) = \sin(x)$ at $x = 1$. Then verify numerically that the error is $O(h^2)$.*

We type the following commands into MATLAB and get the following output for the centered difference calculations and the associated error (note $f'(1) = \cos(1)$)

```
>> h = (1/2).^[1:10];
>> fprime_cd = ( sin(1+h)-sin(1-h))./(2*h);
>> error = fprime_cd - cos(1);
>> [h',fprime_cd',error',[0,error(1:end-1)./error(2:end)]']
```

```
ans =

   0.500000000000000   0.518069447999851  -0.022232857868288                   0
   0.250000000000000   0.534691718664504  -0.005610587203636   3.962661493592264
   0.125000000000000   0.538896367452272  -0.001405938415867   3.990635109130571
   0.062500000000000   0.539950615251025  -0.000351690617114   3.997656882071730
   0.031250000000000   0.540214370333548  -0.000087935534592   3.999414101994002
   0.015625000000000   0.540280321179402  -0.000021984688737   3.999853518155445
   0.007812500000000   0.540296809645632  -0.000005496222508   3.999963376000857
   0.003906250000000   0.540300931809369  -0.000001374058770   3.999990849374081
   0.001953125000000   0.540301962353254  -0.000000343514885   3.999997754440956
   0.000976562500000   0.540302219989371  -0.000000085878769   3.999997802275028
```

As in the previous example, the first column is h, the second is the forward difference approximation, the third column is the error, and the fourth column is the ratios of the successive errors. To verify the method is $O(h^2)$, we expect that the errors get cut in fourth when h gets cut in half, and the fourth column verifies this is true.

4.1.3 Three point difference formulas

The centered difference formula offers a clear advantage in accuracy over the backward and forward difference formulas. However, the centered difference formula cannot be used at the endpoints. Hence if one desires to approximate $f'(x_0)$ or $f'(x_N)$ with accuracy greater than $O(h)$, we have to derive new formulas. The idea in the derivations is to use Taylor series approximations with more points - if we use only two points, we cannot do better than forward or backward difference formulas.

Hence consider deriving a formula for $f'(x_0)$ based on the points $(x_0, f(x_0))$, $(x_1, f(x_1))$, and $(x_2, f(x_2))$. Since we are going to use Taylor series approximations, the obvious choice of the expansion point is x_0. Note that this is the only way to get the equations to contain $f'(x_0)$. For the x points to plug in, we have already decided to use x_0, x_1, x_2, and since we choose x_0 as the expansion point, consider Taylor series for $x = x_1$ and $x = x_2$:

$$f(x_1) = f(x_0) + hf'(x_0) + h^2\frac{f''(x_0)}{2!} + h^3\frac{f'''(x_0)}{3!} + \cdots$$

$$f(x_2) = f(x_0) + 2hf'(x_0) + (2h)^2\frac{f''(x_0)}{2!} + (2h)^3\frac{f'''(x_0)}{3!} + \cdots .$$

The goal is to add scalar multiples of these equations together so that we get a formula in terms of $f(x_0), f(x_1)$ and $f(x_2)$ that is equal to $f'(x_0) + O(h^k)$ where k is as large as possible. The idea is thus to 'kill off' as many terms after the $f'(x_0)$ term as possible. For these two equations, this is achieved by adding $-4\times$ equation 1 + equation 2. This will kill the f'' term, but will leave the f''' term. Hence we truncate the Taylor series' at the f''' term, and get the equation

$$3f(x_0) - 4f(x_1) + f(x_2) = -2hf'(x_0) + \frac{h^3}{6}(-4f'''(c_1) + 8f'''(c_2)),$$

where $x_0 \le c_1 \le x_1$ and $x_0 \le c_2 \le x_2$. Assuming that f is three times differentiable near x_0, dividing through by $-2h$ gives

$$\frac{3f(x_0) - 4f(x_1) + f(x_2)}{-2h} = f'(x_0) + C(h^2),$$

with C being a constant independent of h. A formula for $f'(x_N)$ can be derived analogously, and note that we could have used any three consecutive x-points and got an analogous result. Thus we have proven the following:

Theorem 23. *Let f be three times differentiable near x_i. Then the error in approximating f' with the three point difference approximation satisfies*

$$\left|\frac{-3f(x_i) + 4f(x_{i+1}) - f(x_{i+2})}{2h} - f'(x_i)\right| = O(h^2),$$

$$\left|\frac{3f(x_i) - 4f(x_{i-1}) + f(x_{i-2})}{2h} - f'(x_i)\right| = O(h^2).$$

Thus, if we wanted second order accurate formulas for $f'(x_i)$, we would use the centered difference formula at all interior points, and the three point formulas at the endpoints.

4.1.4 Further notes

There are some important notes to these approximations that should be considered.
-   If you use more points in the formulas, and assume that higher order derivatives of f exist, you can derive higher order accurate formulas. However, even more complicated formulas will be needed at the boundary to get the higher order accuracy. The $O(h^2)$ formulas are by far the most common in practice.
-   If you do not have equal point spacing, $O(h^2)$ formulas can still be derived at each point x_i, using Taylor series as in this section.
-   If you are trying to approximate f' for given data, if the data is noisy, using these methods is probably not a good idea. A better alternative is to find a 'curve of best fit' function for the noisy data, then take the derivative of the function. We discuss curve fitting in a later chapter of this book.

4.2 Approximating the second derivative

Similar ideas as for the first derivative approximations can be used for the second derivative. The procedure for deriving the formulas still use Taylor polynomials, but now we must be careful that our formulas for f'' are only in terms of function values and not in terms of f'. The following is a 3 point centered difference approximation to $f''(x_i)$:

$$f''(x_i) \approx \frac{f(x_{i-1}) - 2f(x_i) + f(x_{i+1})}{h^2}.$$

We can prove this formula is $O(h^2)$.

Theorem 24. *Assume f is four times differentiable in an interval containing x_i. Then*

$$\left| f''(x_i) - \frac{f(x_{i-1}) - 2f(x_i) + f(x_{i+1})}{h^2} \right| = O(h^2).$$

Proof. Let x_i be the expansion point in a Taylor series, and get 2 equations from this series by choosing x values of x_{i+1} and x_{i-1}, and then truncating the series at the fourth derivative terms. This gives us

$$f(x_{i+1}) = f(x_i) + hf'(x_i) + h^2\frac{f''(x_i)}{2!} + h^3\frac{f'''(x_i)}{3!} + h^4\frac{f''''(c_1)}{4!}$$

$$f(x_{i-1}) = f(x_i) - hf'(x_i) + h^2\frac{f''(x_i)}{2!} - h^3\frac{f'''(x_i)}{3!} + h^4\frac{f''''(c_2)}{4!}$$

with $x_{i-1} \le c_2 \le x_i \le c_1 \le x_{i+1}$. Now adding the formulas together and a little algebra provides

$$\frac{f(x_{i-1}) - 2f(x_i) + f(x_{i+1})}{h^2} = f''(x_i) + h^2 \frac{f''''(c_1) + f''''(c_2)}{4!}.$$

The assumption that f is four times differentiable in an interval containing x_i means that for h small enough, the term $\frac{f''''(c_1)+f''''(c_2)}{4!}$ will be bounded above, independent of h. This completes the proof.   □

More accurate (and more complicated) formulas can be derived for approximating the second derivative, by using more x points in the formula. However, we will stop our derivation of formulas here.

4.3 Application: Initial value ODE's using the forward Euler method

Consider the initial value ODE: For a given f and initial data (t_0, y_0), find a function y satisfying

$$y'(t) = f(t, y), \quad y(t_0) = y_0$$

on the interval $[t_0, T]$. Note that this equation can be considered vector or scalar, and is therefore applicable to all initial value problems that can be reduced to a system of first order equations. In a sophomore level differential equations course, students learn how to analytically solve a dozen or so special cases. But how do you solve the other infinitely many cases? Finite difference methods can be an enabling technology for such problems.

We discuss now the forward Euler method for approximating solutions to the ODE above. This is the first, and simplest solver for initial value ODEs, and we will discuss and analyze this and more complex (and more accurate) methods in detail in a later chapter.

We begin by discretizing time with $N + 1$ points, so let

$$t_0 < t_1 < t_2 < t_3 < \cdots < t_N = T$$

be these points. For simplicity, assume the points are equally spaced and call the point spacing Δt. Since we cannot find a function $y(t)$, we wish to approximate the solution $y(t)$ at each t_i; call these approximations y_i. These are the unknowns, and we want them to satisfy $y_i \approx y(t_i)$ in some sense.

Consider our ODE at a particular t, $t = t_n$. Surely the equation holds at t_n since it holds for all t. Thus we have that

$$y'(t_n) = f(t_n, y(t_n)).$$

Applying the forward difference formula gives us

$$\frac{y(t_{n+1}) - y(t_n)}{\Delta t} \approx f(t_n, y(t_n)).$$

The **Forward Euler timestepping algorithm** is created by replacing the approximation signs by equals signs, and the true solutions $y(t_n)$ by their approximations y_n:

Step 1: Given y_0

Steps $n = 1, 2, 3, \ldots, N - 1$: $y_{n+1} = y_n + \Delta t f(t_n, y_n)$.

This is a simple and easy-to-implement algorithm that will yield a set of points

$$(t_0, y_0), (t_1, y_1), \ldots, (t_N, y_N)$$

to approximate the true solution $y(t)$ on $[t_0, T]$. The question of accuracy will be addressed in more detail in a later chapter, but for now we will assume the accuracy is $O(\Delta t)$. We know that the forward difference approximation itself is only first order, so we could not expect forward Euler to be any better than that, but it does reach this accuracy. Hence we can be assured that as we use more and more points (by cutting the timestep Δt), the error will shrink, and the forward Euler solution will converge to the true solution as $\Delta t \to 0$.

The forward Euler code is shown below, which inputs a right hand side function named func, a vector of t's (which discretize the interval $[t_0, T]$), and the initial condition.

```
function y = forwardEuler(func,t,y1)
% y = forwardEuler(func,t,y1)
% solve the ODE y'=f(t,y) with initial condition y(t1)=y1 and return
% the function values as a vector.
% func is the function handle for f(t,y) and t is a vector
% of the times: [t1,t2,...,tn].

% N is the total number of points, initialize y to be the same
% size as t:
N = length(t);
y = zeros(N,1);

% Set initial condition:
y(1)=y1;

% use forward Euler to find y(i+1) using y(i)
for i=1:N-1
    y(i+1) = y(i) + ( t(i+1) - t(i) ) * func(t(i),y(i));
end
```

Example 25. *Given the initial value problem:*

$$y' = \frac{1}{t^2} - \frac{y}{t} - y^2, \quad y(1) = -1,$$

use the forward Euler method to approximate the solution on [1, 2]. *Run it with* 11 *points,* 51 *points, and* 101 *points (so* Δt = 0.1, 0.02, *and* 0.01 *respectively), and compare your solutions to the true solution* $y(t) = \frac{-1}{t}$.

We first need to implement the right hand side $f(t, y) = \frac{1}{t^2} - \frac{y}{t} - y^2$ *(here we wrote it as an inline function). The following code approximates the solution using* 11 *points:*

```
>> f = @(t,y) 1/(t^2) - y/t - y^2;
>> t = linspace(1,2,11);
>> y = forwardEuler(f, t, -1);
>> plot(t,y, 'x-')
```

The plot it produces is shown below:

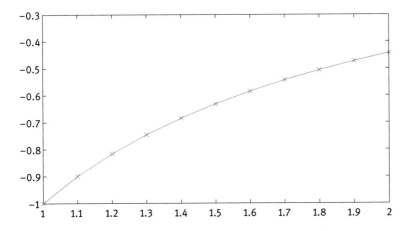

So we have a solution, but is it correct? We now plot the 11 point, 51 point, and 101 point solutions along with the true solution from the command line, as follows:

```
>> f = @(t,y) 1/(t^2) - y/t - y^2;
>>
>> t11 = linspace(1,2,11);
>> y11 = forwardEuler(f, t11, -1);
>>
>> t51 = linspace(1,2,51);
>> y51 = forwardEuler(f, t51, -1);
>>
>> t101 = linspace(1,2,101);
>> y101 = forwardEuler(f, t101, -1);
>>
>> ytrue = -1./t101;
>> plot(t11,y11,'r-',t51,y51,'k--',t101,y101,'b-.',t101,ytrue,'g--', ...
   'LineWidth',1.5)
>> legend('\Delta t=0.1','\Delta t=0.02','\Delta t=0.01','True solution', ...
   'Location', 'NorthWest')
```

which produces the plot below. It is clear to see from the plot that as we use more points (which shrinks the timestep), the forward Euler solution converges to the true solution.

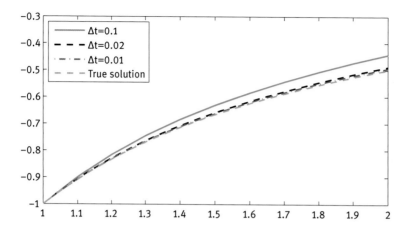

4.4 Application: Boundary value ODE's

Consider next the 1D diffusion equation with given boundary data: Find the function $u(x)$ satisfying

$$u''(x) = f(x) \quad \text{on } a < x < b, \tag{4.7}$$

$$u(a) = u_a, \tag{4.8}$$

$$u(b) = u_b, \tag{4.9}$$

for a given function f and boundary values u_a and u_b for u at the endpoints. Finite difference methods can be used to approximate solutions to equations of this form. Similar to initial value problems, the first step is to discretize the domain, so we pick N equally spaced points on [a,b], with point spacing h:

$$a = x_1 < x_2 < \cdots < x_N = b.$$

Since $u''(x) = f(x)$ on all of the interval (a, b), then it must be true that

$$u''(x_2) = f(x_2),$$
$$u''(x_3) = f(x_3),$$
$$u''(x_4) = f(x_4),$$
$$\cdots\cdots\cdots$$
$$u''(x_{N-1}) = f(x_{N-1}).$$

Now for each equation, we approximate the left hand side using the second order finite difference method. This changes the system of equations to a system of approximations

$$u(x_1) - 2u(x_2) + u(x_3) \approx h^2 f(x_2),$$
$$u(x_2) - 2u(x_3) + u(x_4) \approx h^2 f(x_3),$$
$$u(x_3) - 2u(x_4) + u(x_5) \approx h^2 f(x_4),$$
$$\cdots\cdots\cdots$$
$$u(x_{N-2}) - 2u(x_{N-1}) + u(x_N) \approx h^2 f(x_{N-1}).$$

Denote by u_n the approximation to $u(x_n)$ (for $1 \leq n \leq N$), and let these approximations satisfy the system of approximations above with approximation signs replaced by equals signs. This gives us the system of linear equations for $u_1, u_2, \ldots, u_N$:

$$u_1 - 2u_2 + u_3 = h^2 f(x_2),$$
$$u_2 - 2u_3 + u_4 = h^2 f(x_3),$$
$$u_3 - 2u_4 + u_5 = h^2 f(x_4),$$
$$\cdots\cdots\cdots$$
$$u_{N-2} - 2u_{N-1} + u_N = h^2 f(x_{N-1}).$$

Recall that we know the values of u_1 and u_N since they are given, so we can plug in their values and move them to the right hand sides of their equations. Now the system becomes

$$-2u_2 + u_3 = h^2 f(x_2) - u_a$$
$$u_2 - 2u_3 + u_4 = h^2 f(x_3)$$
$$u_3 - 2u_4 + u_5 = h^2 f(x_4)$$
$$\cdots\cdots\cdots$$
$$u_{N-2} - 2u_{N-1} = h^2 f(x_{N-1}) - u_b.$$

This is a system of $N - 2$ linear equations with $N - 2$ unknowns, which means it can be written in matrix-vector form as $Ax = b$, and we can use Gaussian elimination to solve it.

The same procedure can easily be extended to advection-diffusion-reaction system, i.e. equations of the form:

$$\alpha u''(x) + \beta u'(x) + \gamma u(x) = f(x) \quad \text{on } a < x < b \tag{4.10}$$
$$u(a) = u_a \tag{4.11}$$
$$u(b) = u_b, \tag{4.12}$$

where α, β, γ are constants. Here, we approximate the second derivative in the same way, and for the first derivative, we use the centered difference method. Before applying the boundary conditions, the resulting linear system is

$$\alpha \frac{u_1 - 2u_2 + u_3}{h^2} + \beta \frac{u_3 - u_1}{2h} + \gamma u_2 = f(x_2)$$

$$\alpha \frac{u_2 - 2u_3 + u_4}{h^2} + \beta \frac{u_4 - u_2}{2h} + \gamma u_3 = f(x_3)$$

$$\alpha \frac{u_3 - 2u_4 + u_5}{h^2} + \beta \frac{u_5 - u_3}{2h} + \gamma u_4 = f(x_4)$$

$$\cdots\cdots\cdots$$

$$\alpha \frac{u_{N-2} - 2u_{N-1} + u_N}{h^2} + \beta \frac{u_N - u_{N-2}}{2h} + \gamma u_{N-1} = f(x_{N-1}).$$

Next, using that $u_1 = u_a$ and $u_N = u_b$ and doing some algebra, we get the linear system

$$\left(\frac{-2\alpha}{h^2} + \gamma\right) u_2 + \left(\frac{\alpha}{h^2} + \frac{\beta}{2h}\right) u_3 = f(x_2) - \frac{\alpha u_a}{h^2} + \frac{\beta u_a}{2h}$$

$$\left(\frac{\alpha}{h^2} - \frac{\beta}{2h}\right) u_2 + \left(\frac{-2\alpha}{h^2} + \gamma\right) u_3 + \left(\frac{\alpha}{h^2} + \frac{\beta}{2h}\right) u_4 = f(x_3)$$

$$\left(\frac{\alpha}{h^2} - \frac{\beta}{2h}\right) u_3 + \left(\frac{-2\alpha}{h^2} + \gamma\right) u_4 + \left(\frac{\alpha}{h^2} + \frac{\beta}{2h}\right) u_5 = f(x_4)$$

$$\cdots\cdots\cdots$$

$$\left(\frac{\alpha}{h^2} - \frac{\beta}{2h}\right) u_{N-2} + \left(\frac{-2\alpha}{h^2} + \gamma\right) u_{N-1} = f(x_{N-1}) - \frac{\alpha u_b}{h^2} - \frac{\beta u_b}{2h}.$$

Notice that the coefficients in each equation form a pattern, and this can be exploited to easily generate the resulting matrix. We give the code below for solving this problem for general α, β, γ, f, u_a and u_b. The user inputs each of these, as well as the total number of points to use in the discretization.

```
function [x,u] = convDiffReact(f,a,b,N,ua,ub,alpha,beta,gamma)
% function [x,u] = convDiffReact(f,a,b,N,ua,ub,alpha,beta,gamma)
% input: function f, interval [a,b], N is the number of points


    % discretize the interval, h is the point spacing
    x = linspace(a,b,N)';
    h = (b-a)/(N-1);

    % setup matrix for u'' (will have N-2 unknowns)
    e = ones(N-2,1);
    A = spdiags([(alpha/(h^2) - beta/(2*h))*e, ...
        (-2*alpha/(h^2) + gamma)*e,...
        (alpha/(h^2)+ beta/(2*h))*e], -1:1, N-2, N-2);

    % setup rhs vector
    b = feval(f,x(2:end-1));
```

```
b(1)  =  b(1)-alpha*ua/(h^2)  +  beta*ua/(2*h);
b(end)=b(end)-alpha*ub/(h^2)  -  beta*ub/(2*h);

% solve the linear system
utemp = A \ b;

% put solution together to include boundary points, and so it
% correponds to x's at same index
u = [ua;utemp;ub];
```

We now present an example.

Example 26. *Use the finite difference method to solve the boundary value problem*

$$y'' = 2y' - y + xe^x - x, \ 0 < x < 2, \ y(0) = 0, \ y(2) = -4.$$

The true solution is given by $y_{true} = \frac{1}{6}x^3 e^x - \frac{5}{3}xe^x + 2e^x - x - 2$. Use this to verify the code is correct, and converges to the true solution with rate $O(h^2)$.

Matching this to our general model problem, we get that $\alpha = 1$, $\beta = -2$, $\gamma = 1$ and $f(x) = xe^x - x$. Hence we can use our code to solve it, after we write the function for the right hand side function f:

```
function f = CDRfun(x)
    f = x.*exp(x) - x;
```

We can now use MATLAB to solve the problem. The commands to do this, and plot the solution are:

```
>> [x,u] = convDiffReact(@CDRfun,0,2,100,0,-4,1,-2,1);
>> plot(x,u)
```

and the plot we get is

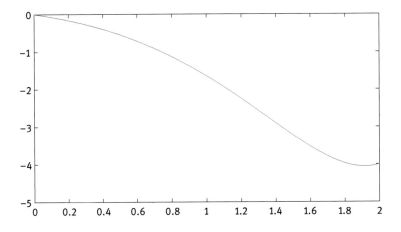

A fundamental question is whether this approximation is accurate. If we did not know the true solution, the strategy would be to run several successively finer meshes (e.g. N = 200, 400, 800), and when the solution converges then we can usually believe it.

In this case, we know the true solution, and so we verify the accuracy of the solution (and thus of the code too) by comparing to the true solution. If the method is $O(h^2)$, then as we cut h in half, we expect that the error gets cut in fourth. For 'error' here, we use the maximum error at any point, compared to the true solution. We use h = 0.01, 0.005, 0.0025 and 0.00125, that is, we keep cutting h in half and rerunning. The MATLAB commands to do this are below, as well as the errors. We observe from the output that the errors get cut by approximately 4 as we cut h in half, which verifies the method is $O(h^2)$.

```
>> [x,u] = convDiffReact(@CDRfun,0,2,101,0,-4,1,-2,1);
>> utrue = 1/6*(x.^3).*exp(x) - 5/3*x.*exp(x) + 2*exp(x) - x - 2;
>> max( abs(u-utrue))

ans =

    9.5947e-05

>> [x,u] = convDiffReact(@CDRfun,0,2,201,0,-4,1,-2,1);
>> utrue = 1/6*(x.^3).*exp(x) - 5/3*x.*exp(x) + 2*exp(x) - x - 2;
>> max( abs(u-utrue))

ans =

    2.3979e-05

>> [x,u] = convDiffReact(@CDRfun,0,2,401,0,-4,1,-2,1);
>> utrue = 1/6*(x.^3).*exp(x) - 5/3*x.*exp(x) + 2*exp(x) - x - 2;
>> max( abs(u-utrue))

ans =

    5.9943e-06

>> [x,u] = convDiffReact(@CDRfun,0,2,801,0,-4,1,-2,1);
>> utrue = 1/6*(x.^3).*exp(x) - 5/3*x.*exp(x) + 2*exp(x) - x - 2;
>> max( abs(u-utrue))

ans =

    1.4986e-06
```

4.5 Exercises

1. Find Taylor series approximations using quadratic polynomials ($n = 2$) for the function $f(x) = e^x$ at $x = 0.9$, using the expansion point $x_0 = 1$. Find an upper bound on the error using Taylor's theorem, and compare it to the actual error.

2. Prove Theorem 18.

3. Approximate the derivative of $f(x) = \cos^2(x)$ at $x = 1$, using backward, forward and centered difference approximations. Use the same h as in the examples in this section. Verify numerically the error (the convergence order) is $O(h)$, $O(h)$ and $O(h^2)$, respectively.

4. Show that
$$\left| \frac{f(x_{n+3}) - 9f(x_{n+1}) + 8f(x_n)}{-6h} - f'(x_n) \right| = O(h^2).$$

5. The following data represents (time (s),distance traveled (ft)) for a car:
    $(0,0)$, $(2,224)$, $(4,384)$, $(6,625)$, $(8,746)$, $(10,994)$.
    Use second order formulas to approximate the speed in miles per hour at $t = 0$, $t = 8$, and $t = 10$.

6. Use the finite difference method to approximate a solution to the boundary value problem
$$y'' - 5y' - 2y = x^2 e^x, \quad -1 < x < 2, \quad y(-1) = 0, \quad y(2) = 0.$$
    Run the code using $N = 11, 21, 51, 101, 201$, and 401 points. Plot all the solutions on the same graph. Has it converged by $N = 401$?

7. Water flows into a cylindrical tank from an open top, and out of the tank at the bottom through a pipe. A differential equation for the water height, h, in a specific problem setting, is given in meters by
$$3.2h'(t) = 0.3 + 0.2\cos(\pi t/12) - 0.06\sqrt{19.6h}.$$
    If the initial water height is 3m, find the height of the water for $0 \le t \le 200$. Use the forward Euler method to approximate a solution.
    Include a plot of the solution with your answer, and also discuss why you think your answer is correct.

8. The differential equation
$$W''(x) - \frac{S}{D}W(x) = \frac{-ql}{2D}x + \frac{q}{2D}x^2, \quad 0 \le x \le l$$
    describes plate deflection under an axial tension force, where l is the plate length, D is rigidity, q is intensity of the load, and S is axial force. Approximate the deflection of a plate if $q = 200 \text{ lb/in}^2$, $S = 200 \text{ lb/in}$, $D = 1,000,000 \text{ lb/in}$ and $l = 12 \text{ in}$, assuming $W(0) = W(l) = 0$. Plot your solution, and discuss why you think it is correct.

5 Solving nonlinear equations

There should be no doubt to anyone that there is a great need to solve nonlinear equations with numerical methods. Just in one variable, we have equations such as

- $e^x = x^2$,
- $\sin(x) = x + 1$,
- $\sin^3(x) + 2\sin^2(x) + \cos(x) = 5e^{\cos(x)}$,

which are difficult or impossible to solve analytically. In multiple variables, it only gets worse. In a later chapter on numerical integration, we will want a solution (x_1, x_2, w_1, w_2) to the system

$$w_1 + w_2 = 2, \tag{5.1}$$
$$w_1 x_1 + w_2 x_2 = 0, \tag{5.2}$$
$$w_1 x_1^2 + w_2 x_2^2 = \frac{2}{3}, \tag{5.3}$$
$$w_1 x_1^3 + w_2 x_2^3 = 0. \tag{5.4}$$

For most nonlinear equations, finding an analytical solution is impossible or very difficult. The purpose of this chapter is to study some common numerical methods for solving nonlinear equations: we will quantify how they work, when they work, how well they work, and when they fail.

We note that solving nonlinear equations is equivalent to root-finding for a function f. For any system of equations, if we simply move all the terms to the left hand side(s) of the equation(s), then the problem in general we wish to solve take the forms

$$\text{1 variable:} \qquad f(x) = 0$$
$$\text{n variables:} \qquad \mathbf{f}(\mathbf{x}) = \mathbf{0},$$

where $\mathbf{x} = (x_1, x_2, \ldots, x_n)$ and $\mathbf{f} = \langle f_1, f_2, \ldots, f_n \rangle$, and thus $\mathbf{f}(\mathbf{x}) = \mathbf{0}$ is just the vector representation of the system of n nonlinear equations in n variables. For example, in the system above, we can define

$$f_1(x_1, x_2, w_1, w_2) = w_1 + w_2 - 2$$
$$f_2(x_1, x_2, w_1, w_2) = w_1 x_1 + w_2 x_2,$$
$$f_3(x_1, x_2, w_1, w_2) = w_1 x_1^2 + w_2 x_2^2 - \frac{2}{3},$$
$$f_4(x_1, x_2, w_1, w_2) = w_1 x_1^3 + w_2 x_2^3,$$

and then would find a root of $\mathbf{f} = \langle f_1, f_2, f_3, f_4 \rangle$.

5.1 The bisection method

The bisection method is very robust method for solving nonlinear equations in 1 variable. It is based on a special case of the intermediate value theorem, which we recall now from Calculus:

Theorem 27 (Intermediate value theorem). *Suppose a function f is continuous on [a,b], and $f(a) \cdot f(b) < 0$. Then there exists a number c in the interval (a, b) such that $f(c) = 0$.*

The theorem says that if $f(a)$ and $f(b)$ are of opposite signs, and f is continuous, then the graph of f must cross $y = 0$ between a and b, and call c the x-point where it crosses. Consider the following illustration of the theorem. In the picture below, two points are shown whose y-values have opposite signs. Do you think you can draw a continuous curve that connects these points without crossing the dashed line? No, you cannot! This is precisely what the intermediate value theorem says.

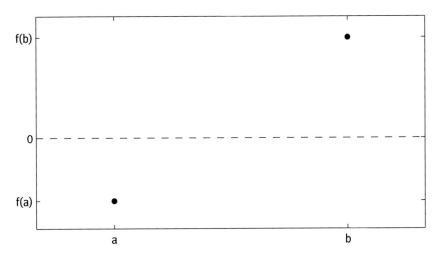

Now that we have established that any continuous curve that connects the points must cross the dashed line, let's look at an example in the picture below. Note that a curve may cross multiple times, but we are guaranteed that it crosses at least once.

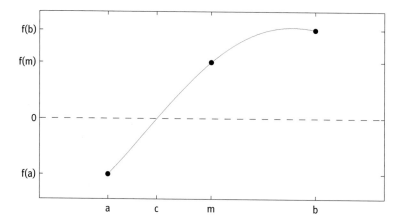

The bisection method for rootfinding is based on a simple idea and uses the intermediate value theorem. Consider two points, $(a, f(a))$, and $(b, f(b))$ from a function/graph with $f(b)$ and $f(a)$ having opposite signs. Thus we know that there is a root of f in the interval (a,b). Let m be the midpoint of the interval [a,b], so $m = (a + b)/2$, and evaluate $f(m)$. One of the following is true: Either $f(m) = 0$, $f(a)$ and $f(m)$ have opposite signs, or $f(m)$ and $f(b)$ have opposite signs - since we know $f(m)$, it is simple to check which one. This means that we now know an interval that the root lives in **that is half the width of the previous interval!** So if we repeat the process over and over, we can repeatedly reduce the size of the interval which we know contains a root. Do this enough times to "close in" on the root, and then the midpoint of that final interval can be considered a good estimate of a root.

Let us now illustrate this process.

Example 28. *Use the bisection method to find the root of $f(x) = \cos(x) - e^x$ on $[-2, -0.5]$.*

Clearly the function is continuous on the interval, and now we check the function values at the endpoints have opposite signs:

$$f(-2) = -0.551482119783755 < 0 \qquad f(-0.5) = 0.271051902177739 > 0.$$

Hence by the intermediate value theorem, we know that a root exists in $[-2, -0.5]$.

Call m the midpoint of the interval, so $m = (-2 + (-0.5))/2 = -1.25$, and we evaluate $f(-1.25) = 0.028817565535079 > 0$. Thus there is a sign change on $[-2, -1.25]$, so we now know an interval containing a root is $[-2, -1.25]$ since there is a sign change in the function values! Hence we have reduced in half the width of the interval.

For our next step, call m the midpoint of $[-2, -1.25]$, so $m = -1.625$ and then we calculate $f(-1.625) = -0.251088810231130 < 0$. Thus there is a sign change in $[-1.625, -1.25]$, so we keep this interval and move on.

For our next step, call m the midpoint of $[-1.625, -1.25]$, so $m = -1.4375$ and then we calculate $f(-1.4375) = -0.104618874642633 < 0$. Thus there is a sign change in $[-1.4375, -1.25]$, so we keep this interval and move on.

Repeating this process over and over again will zoom in on the root −1.292695719373398. *Note that, if the process is terminated after some number of iterations, we will know the error between the last midpoint and true solution is half the length of the last interval.*

Clearly, this process repeats itself on each step. We give below a code for the bisection process.

```
function [y,data] = bisect(a,b,func,tol)
% bisect(a,b,func,tol), uses the bisection method to find a root of
% func in (a,b) within tolerance tol

% evaluate the function at the end points
fa = func(a);
fb = func(b);

% Check that fa and fb have opposite signs (since we implicitly
% assume this)
if fa*fb >= 0
    error('The function fails to satisfy f(a)*f(b)<0 for inputs' ...
        'a and b')
end

% keep track of progress of algorithm
data = [0 a b fa fb b-a];

% now we can run the algorithm
it_count = 0;
while (b-a) > tol
    it_count = it_count+1;
    % Get the midpoint and evaluate it
    xnew = (a+b)/2;
    fnew = func(xnew);

    % determine which interval to keep
    if fa * fnew <= 0
        b  = xnew;
        fb = fnew;
    else
        a  = xnew;
        fa = fnew;
    end

    datanew = [it_count a b fa fb b-a];
    data = [data;datanew];
end
```

```
% The solution needs to be a number - choose the midpoint of the
% final interval
y = (a+b)/2;

end
```

Now let's use the code to solve a nonlinear equation.

Example 29. *Find a solution to* $e^x = x^2$ *using the bisection method to within a tolerance of* 10^{-6}.

The bisection method finds 0's of functions, so first we note that we will try to find a zero of the function $f(x) = e^x - x^2$. For this, we define the function (here as an inline function):

```
>> myfun1 = @(x) exp(x)-x^2;
```

Now we use our bisection function. Since $f(0) = 1 > 0$ and $f(-1) = e^{-1} - 1 < 0$, we are guaranteed that bisection will find a root on this interval. So now run it:

```
>> [root,data] = bisect(-1,0,myfun1,1e-6);
```

We can print out the root:

```
>> root

root =

  -7.0347e-01
```

and check that it really is a zero of the function:

```
>> myfun1(root)

ans =

  -8.0526e-07
```

Our code kept track of the data a, b, f(a), f(b), and b-a at each iteration. We can view this as well:

```
>> data

data =

         0  -1.0000e+00          0  -6.3212e-01   1.0000e+00   1.0000e+00
1.0000e+00  -1.0000e+00  -5.0000e-01  -6.3212e-01   3.5653e-01   5.0000e-01
2.0000e+00  -7.5000e-01  -5.0000e-01  -9.0133e-02   3.5653e-01   2.5000e-01
3.0000e+00  -7.5000e-01  -6.2500e-01  -9.0133e-02   1.4464e-01   1.2500e-01
4.0000e+00  -7.5000e-01  -6.8750e-01  -9.0133e-02   3.0175e-02   6.2500e-02
5.0000e+00  -7.1875e-01  -6.8750e-01  -2.9240e-02   3.0175e-02   3.1250e-02
6.0000e+00  -7.1875e-01  -7.0312e-01  -2.9240e-02   6.5113e-04   1.5625e-02
7.0000e+00  -7.1094e-01  -7.0312e-01  -1.4249e-02   6.5113e-04   7.8125e-03
8.0000e+00  -7.0703e-01  -7.0312e-01  -6.7873e-03   6.5113e-04   3.9062e-03
```

```
9.0000e+00   -7.0508e-01   -7.0312e-01   -3.0652e-03   6.5113e-04   1.9531e-03
1.0000e+01   -7.0410e-01   -7.0312e-01   -1.2063e-03   6.5113e-04   9.7656e-04
1.1000e+01   -7.0361e-01   -7.0312e-01   -2.7741e-04   6.5113e-04   4.8828e-04
1.2000e+01   -7.0361e-01   -7.0337e-01   -2.7741e-04   1.8691e-04   2.4414e-04
1.3000e+01   -7.0349e-01   -7.0337e-01   -4.5241e-05   1.8691e-04   1.2207e-04
1.4000e+01   -7.0349e-01   -7.0343e-01   -4.5241e-05   7.0835e-05   6.1035e-05
1.5000e+01   -7.0349e-01   -7.0346e-01   -4.5241e-05   1.2797e-05   3.0518e-05
1.6000e+01   -7.0348e-01   -7.0346e-01   -1.6222e-05   1.2797e-05   1.5259e-05
1.7000e+01   -7.0347e-01   -7.0346e-01   -1.7121e-06   1.2797e-05   7.6294e-06
1.8000e+01   -7.0347e-01   -7.0346e-01   -1.7121e-06   5.5427e-06   3.8147e-06
1.9000e+01   -7.0347e-01   -7.0347e-01   -1.7121e-06   1.9153e-06   1.9073e-06
2.0000e+01   -7.0347e-01   -7.0347e-01   -1.7121e-06   1.0159e-07   9.5367e-07
```

We observe that it took 20 iterations for the interval to shrink smaller than 10^{-6}.

5.2 Newton's method

The next method we study is Newton's method. While the bisection method has the attractive property that it will converge if the method is continuous and has a sign change, it has a disadvantage in that it is not smart enough to take advantage of simple functions. For example, if we know a function is linear, then finding the root should take just one step. But we also know that if you 'zoom in' on a smooth function, it looks linear. Newton figured out how to take advantage of this idea and create a faster root finding algorithm.

Newton's idea is simple: To find the root of a function, find the tangent line of the function at your current guess for the root, then your next guess will be where the tangent line hits the x-axis. Note that if the function is linear, then the root will be found in just one step, and if the function is close to linear, then this will give a very good approximation of the root. To start the algorithm, we need an initial guess, and the better it is, the better the algorithm will work. The idea is illustrated in the following plots:

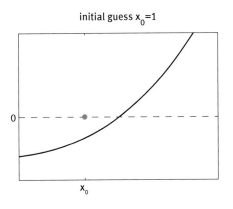

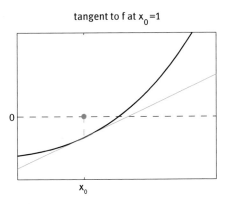

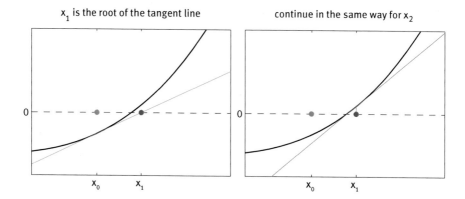

If we know the previous guess x_k, Newton's idea allows us to explicitly find x_{k+1}. The tangent line to f at x_k is given by

$$y - f(x_k) = f'(x_k)(x - x_k).$$

The next Newton iteration is when this line crosses the x-axis. Thus the point where it crosses the axis is $(x_{k+1}, 0)$. Plugging this into the tangent line gives

$$0 - f(x_k) = f'(x_k)(x_{k+1} - x_k).$$

Now solve for x_{k+1}:

$$x_{k+1} = x_k - \frac{f(x_k)}{f'(x_k)}.$$

Thus given an initial guess x_0, the formula above can be used to calculate x_1, x_2, and so on, as far as you like. We would stop once $|x_{k+1} - x_k| < tol$ for some specified tolerance tol. Alternative stopping criteria are $f(x_k) < tol$ (check error) or $\frac{x_{k+1} - x_k}{x_k} < tol$ (relative tolerance). Thus we have defined another algorithm for rootfinding:

Algorithm 30 (Newton's method).

Given: f, f', tol, x_0
while($|x_{k+1} - x_k| > tol$):
$$x_{k+1} = x_k - \frac{f(x_k)}{f'(x_k)}$$

The code is simple to write:

```
function [root,numits] = newt(fun,funderiv,x0,tol)
% solve the equation fun(x)=0 using Newton's method with starting
% value x0 up to the tolerance tol. 'funderiv' is the derivative of
% 'fun'

xold=x0+1; % initialize to something other than x0 to enter loop
xnew=x0;
numits = 0;
```

```
while ( abs( xnew-xold ) > tol )
    % perform the Newton iteration
    fxnew = fun(xnew);
    fprimexnew = funderiv(xnew);
    numits = numits+1;

    xold = xnew;
    xnew = xnew - fxnew/fprimexnew;
end

root = xnew;
end
```

Example 31. *Use Newton's method to solve* $e^x = x^2$ *using initial guess* $x_0 = -1$ *and tolerance* 10^{-14}.

We need to find the zero of $f(x) = e^x - x^2$ *so we define* $f(x)$ *(like in the last section) and the derivative* $f'(x)$ *so we can pass them to* newt:

```
>> myfun1 = @ (x) exp(x) - x^2;
>> myfun1prime = @(x) exp(x) - 2*x;
>> [root,numiterations]=newt(myfun1,myfun1prime,-1,1e-14)
root =
   -0.703467422498392
numiterations =
     5
```

We observe that the correct answer is found to a much smaller tolerance than the bisection method in only 5 iterations! It is typical for Newton's method to converge much faster than bisection, and we quantify this below.

While the bisection method always converges if the function is continuous and the starting interval contains a root, Newton's method can run into one of several difficulties:

1. Vanishing derivatives. If $f'(x_k) \approx 0$ the tangent line has no zero or one that is very far away. Then Newton's method becomes numerically unstable and will not converge.
2. Cycles. You can construct examples where Newton's method jumps between two points (so $x_{k+2} = x_k$) and never converges. One example is $f(x) = x^3 - 2x + 2$ with $x_1 = 0$.

In summary, Newton's method is not guaranteed to work. It works as long as the starting value is "close enough" to the root.

5.3 Secant method

Although Newton's method is much faster than bisection, it has a disadvantage in that it explicitly uses the derivative of the function. In many processes where rootfinding is needed (optimization processes for example), we may not know an analytical expression for the function, and thus may not be able to find an expression for the derivative. That is, a function evaluation may be the result of a process or experiment. While we may theoretically expect the function to be differentiable, we cannot explicitly get the derivative. For situations like this, the secant method has been developed, and one can think of the secant method as being the same as Newton's method, except that instead of using the tangent line at x_k, you use the secant line at x_k and x_{k-1}. This defines the following algorithm:

Algorithm 32 (Secant method).
Given: f, tol, x_0, x_1
while($|x_{k+1} - x_k| > tol$):

$$x_{k+1} = x_k - \frac{f(x_k)}{\frac{f(x_k)-f(x_{k-1})}{x_k-x_{k-1}}}$$

Hence we may think of the secant method as Newton's method, but with the derivative term $f'(x_k)$ replaced by the backward difference $\frac{f(x_k)-f(x_{k-1})}{x_k-x_{k-1}}$.

5.4 Comparing bisection, Newton, secant method

We will now quantify how fast the three methods converge to the solution. For this we will define:

Definition 33. *Let e_n be the error in step n of an algorithm. We say an iteration converges with order k if*

$$\lim_{n\to\infty} \frac{|e_{n+1}|}{|e_n|^k} = C < \infty,$$

with $C < 1$ if $k = 1$. If $k = 1$ we call the convergence "linear", if $k = 2$ "quadratic".

We have seen before that the bisection method always halves the size of the interval in every step. If we define the error in step n as the size of the interval, then the bisection method is of order 1 because $e_{n+1} = \frac{1}{2}e_n$. Note that we are using the interval length (an upper bound on the error) as the error and not the actual error for the interval midpoint. The reason is for convenient error quantification since we know precisely the behavior of the interval width at each step, but the behavior of the real error can be erratic.

We state the result for Newton's method without proof.

Theorem 34. *Let f be twice continuously differentiable around a root x^*, and let $f'(x^*) \neq 0$. Then Newton's method convergences quadratically for all x_0 sufficiently close to x^*.*

The secant method converges (again with some technical conditions on f) with order $k \approx 1.618$ (the golden ratio).

Let us now summarize the properties of the three methods:

	Bisection	**Newton**	**Secant**
requirements on f?	continuous, sign change in $[a, b]$	continuous, differentiable	continuous, differentiable
needs	interval $[a, b]$	point x_0	points x_0 and x_1
always converges?	yes	no, only if "close"	no, only if "close"
higher dimensions?	no	yes	yes[1]
work per step	1 function	1 function, 1 derivative[2]	1 function
convergence rate[3]	1	2	~1.6

[1] The secant method can be extended to higher dimensions, but the approximation of the derivative requires more work. There are several similar methods known as "Quasi-Newton" or "Jacobian-free Newton" methods.

[2] The user needs to supply the derivative to the algorithm, which can be problematic, for example if it is difficult to compute by hand.

[3] To achieve the given convergence rate there are more technical requirements on f (smoothness) and it only works if starting "close enough" (see above). Nevertheless, Newton's is typically faster than secant, which is in turn faster than bisection.

5.5 Combining secant and bisection and the *fzero* command

From the comparisons, we see that bisection is the most robust (least number of assumptions to get guaranteed convergence), but is slow. On the other hand, secant is much faster but needs good initial guesses. Hence is seems reasonable that an algorithm that used bisection method to get close to a root, then the secant method to 'zoom in' quickly, would get the best of both worlds. Indeed, this can be done with a simple algorithm: Start with an interval that his a sign change, and then instead of using the midpoint to shrink the interval, use the secant method. But if the secant method gives a guess that is outside of the interval, then use the midpoint instead. Eventually, the secant method will be used at each step.

MATLAB/Octave have a function built-in that does this (and more) for functions in one variable, and it is called 'fzero'. To use it, you simply need to give it a user-defined function and an initial guess or interval. The function intelligently alternates between bisection, secant, and a third method called inverse quadratic interpolation (use 3 points to fit a quadratic and use a root - if one exists - as the next guess).

Consider now using 'fzero' to solve $x^3 = \sin(x) + 1$. First, make a function where the solution of the equation is a root of the function, as in the following:

```
function y = myNLfun1(x)
    y=x^3 - sin(x) - 1;
```

Then simply pass this function into fzero, and give it an initial guess:

```
>> fzero(@myNLfun1,5)

ans =

    1.249052148501195

>> myNLfun1(1.249052148501195)

ans =

    1.332267629550188e-15
```

Options are available in MATLAB to set tolerances, but as you see above, often the default provides a very accurate answer. You can see what fzero is doing by specifying some additional parameters:

```
>> x=fzero(fun,5,optimset('Display','iter'));
```

Search for an interval around 5 containing a sign change:

Func-count	a	f(a)	b	f(b)	Procedure
1	5	124.959	5	124.959	initial interval
3	4.85858	114.68	5.14142	135.819	search
5	4.8	110.588	5.2	140.491	search
7	4.71716	104.964	5.28284	147.277	search
9	4.6	97.3297	5.4	157.237	search
11	4.43431	87.1542	5.56569	172.065	search
13	4.2	73.9596	5.8	194.577	search
15	3.86863	57.5637	6.13137	229.652	search
17	3.4	38.5595	6.6	286.184	search
19	2.73726	19.1157	7.26274	381.261	search
21	1.8	3.85815	8.2	549.427	search
22	0.474517	-1.35006	8.2	549.427	search

Search for a zero in the interval [0.474517, 8.2]:

Func-count	x	f(x)	Procedure
22	0.474517	-1.35006	initial
23	0.493453	-1.35352	interpolation
24	0.493453	-1.35352	bisection
25	0.555978	-1.35592	interpolation
26	0.555978	-1.35592	bisection
27	0.733834	-1.27454	interpolation
28	0.733834	-1.27454	bisection
29	1.06409	-0.669488	interpolation
30	1.33921	0.428557	interpolation
31	1.23183	-0.0738933	interpolation
32	1.24763	-0.00621393	interpolation
33	1.24906	1.46004e-05	interpolation

```
34          1.24905  -2.01574e-08      interpolation
35          1.24905   -6.4837e-14      interpolation
36          1.24905            0       interpolation
```
```
Zero found in the interval [0.474517, 8.2]
```

5.6 Equation solving in higher dimensions

Without getting too deep into details, we simply state here that the Newton algorithm works in higher dimensions, with the same 'good' and 'bad' points as in the 1D case. The only difference in the algorithm is that instead of

$$x_{k+1} = x_k - \frac{f(x_k)}{f'(x_k)}$$

we have to use

$$\mathbf{x}_{k+1} = \mathbf{x}_k - (\nabla \mathbf{f}(\mathbf{x}_k))^{-1} \mathbf{f}(\mathbf{x}_k),$$

where in the 2×2 case,

$$\mathbf{f}(\mathbf{x}) = (f_1(x_1, x_2), f_2(x_1, x_2)).$$

and

$$\nabla \mathbf{f} = \begin{pmatrix} \dfrac{\partial f_1}{\partial x_1} & \dfrac{\partial f_1}{\partial x_2} \\ \dfrac{\partial f_2}{\partial x_1} & \dfrac{\partial f_2}{\partial x_2} \end{pmatrix}.$$

As an example, consider solving the system of nonlinear equations,

$$x_1^3 = x_2,$$
$$x_1 + \sin(x_2) = -3.$$

First we create MATLAB functions for the function and its derivative (put these into separate .m files):

```
function y = myNDfun( x )
y(1,1) = x(1)^3 - x(2);
y(2,1) = x(1) + sin(x(2)) + 3;
end
```

```
function y = myNDfunprime( x )
y(1,1) = 3*x(1)^2;
y(1,2) = -1;
y(2,1) = 1;
y(2,2) = cos(x(2));
end
```

Next, create the *n* dimensional Newton method function:

```
function [root,numits] = newton(fun,gradfun,x0,tol)
% Solve fun(x)=0 using Newton's method given the function and its
% gradient gradfun starting from the initial guess x0.

x0 = x0(:); % this will force x0 to be a column vector
xold = x0+1;% this needs to be ~= x0 so that we enter the while loop
xnew = x0;
numits = 0;

n = length(x0);

while norm(xnew-xold)>tol

    gradfxk = gradfun(xnew);
    fxk = fun(xnew);
    fxk = fxk(:); % this will force fxk to be a column vector

    [a,b]=size(fxk);
    if a~=n || b~=1
        error('function has wrong dimension, expecting %d x 1,' ...
              'but got %d x %d',n , a, b)
    end
    [a,b]=size(gradfxk);
    if a~=n || b~=n
        error('gradient has wrong dimension, expecting %d x %d,'...
              'but got %d x %d', n, n, a, b)
    end

    xold = xnew;
    % x_k+1 = x_k - (grad f(xk))^{-1} * f(xk), but implement
    % as a linear solve
    xnew = xnew - gradfxk \ fxk;

    numits = numits+1;
    if (numits>=100)
        root = xnew;
        fprintf('current step:\n')
        disp(xnew)
        error('no convergence after %d iterations', numits);
    end
end

root = xnew;
end
```

Running it gives the following answer:

```
>> [root,numits] = newton(@myNDfun,@myNDfunprime,[-2;-15],1e-8)
root =
  -2.474670119857577
 -15.154860516817051
numits =
     5
```

We can check that we are very close to a root:

```
>> value = myNDfun(root)
value =
   1.0e-14 *
  -0.532907051820075
  -0.088817841970013
```

If we start too far away from the root, the algorithm will not converge:

```
>> [root,numits] = newton(@myNDfun,@myNDfunprime,[100;-10],1e-8)
current step:
  -1.990536027147847
  -7.886133867955427
Error using newton (line 36)
no convergence after 100 iterations
Error in ch4_newtondimex3 (line 3)
[root,numits] = newton(@myNDfun,@myNDfunprime,[100;-10],1e-8)
```

5.7 Exercises

1. Determine the root of $f(x) = x^2 - e^{-x}$ by hand using (a) bisection method with $a = 0$ and $b = 1$ and (b) Newton's method with $x_1 = 0$. Carry out three iterations each. You can round the numbers to 5 digits and use a calculator or MATLAB to do each individual calculation.

2. Since the bisection method is known to cut the interval containing the root in half at each iteration, once we are given the stopping criteria "tol" (i.e. stop when $|b - a| <$ tol), we immediately know how many steps of bisection need to be taken. Change the code "bisect.m" to use a for loop instead of a while loop, and so that the answer is to within tol, but the minimum number of bisection iterations are used. You will need to calculate the number of loop iterations.

3. Consider the trisection method, which is analogous to bisection except that at each iteration, it creates 3 intervals instead of 2, and keeps one interval where there is a sign change.
    (a) Is the trisection method guaranteed to converge if the initial interval has a sign change? Why or why not?

(b) Adapt the bisection code to create a code that performs the trisection method. Compare the results of bisection to trisection for "myfun1.m" for the same stopping tolerance, but several different starting intervals. How do the methods compare?

(c) The main computational cost of bisection and trisection is function evaluations. The other operations it does is a few additions and divisions, but the evaluation of most functions, including sin, cos, exp, etc. are much more costly. Based on this information, explain why you would never want to use trisection over bisection.

4. Use the bisection, Newton, and secant methods to find the smallest positive solution to $\sin(x) = \cos(x)$, using a tolerance of 10^{-12}. Note you will create your own secant code - it is a small change from the Newton code in this chapter. Report the solutions, the number of iterations each method needed, the number of function evaluations performed, and the commands you used to find the solutions. Conclude which method is fastest.

5. Which of Newton, secant, and bisection is most appropriate to find the smallest positive root of the function $f(x) = |\sin(100x)| - 1/2$, and why?

6. Use the `fzero` command to find a solution to $\sin^2(x) = \cos^3(x)$.

7. Use Newton's method with initial guess $\langle 0, 0, 0 \rangle$ to find the solution of

$$x_1^2 + x_2 - 33 = 0$$
$$x_1 - x_2^2 - 4 = 0$$
$$x_1 + x_2 + x_3 - 2 = 0$$

to 10 digits of accuracy.

8. Solve the following nonlinear system using Newton's method in MATLAB:

$$x^2 = y + \sin(z)$$
$$x + 20 = y - \sin(10y)$$
$$(1 - x)z = 2.$$

Hint: You need to find a suitable starting value for x, y, and z so that the method converges.

6 Accuracy in solving linear systems

In several applications, we have solved linear systems in order to obtain answers to various problems. An important question which we have not yet considered is whether the solutions are accurate. Some systems are very sensitive to small changes in data or roundoff error, and thus their answers are possibly not good. Other systems are not sensitive, and their solutions are likely very good. In this chapter we will quantify the sensitivity of systems with the notion of matrix conditioning. To do so, we begin with the concept of matrix inverses.

6.1 Gauss–Jordan elimination and finding matrix inverses

A simple variation of Gaussian elimination can be made as follows: instead of using the pivot to zero out just the entries below the diagonal, use it to also zero out entries above the diagonal. From this procedure we will get a diagonal matrix when complete, instead of a triangular one. If we rescale the diagonal entries to be 1, then we will get the identity matrix. This variation of Gaussian elimination is called Gauss-Jordan elimination.

Gauss-Jordan elimination generally should not be used to solve linear systems, since it requires about twice the work of Gaussian elimination. However, Gauss-Jordan elimination is a convenient choice when solving several linear systems simultaneously, since it easily allows each solve to be done at the same time. Still, it is instructive to see it in action, so let us consider using it to solve a linear system in an example.

Example 35 (Gauss-Jordan elimination). *Use Gauss-Jordan elimination to solve the system of equations*

$$\begin{pmatrix} 3 & 1 & 0 \\ 1 & -1 & 1 \\ 0 & 1 & 2 \end{pmatrix} \begin{pmatrix} x_1 \\ x_2 \\ x_3 \end{pmatrix} = \begin{pmatrix} 1 \\ 1 \\ 1 \end{pmatrix}.$$

The pivot in the first column is $a_{11} = 3$, so divide row 1 by 3 to get

$$\begin{pmatrix} 1 & 1/3 & 0 \\ 1 & -1 & 1 \\ 0 & 1 & 2 \end{pmatrix} \begin{pmatrix} x_1 \\ x_2 \\ x_3 \end{pmatrix} = \begin{pmatrix} 1/3 \\ 1 \\ 1 \end{pmatrix}.$$

Now subtract row 1 from row 2 to zero out a_{21}:

$$\begin{pmatrix} 1 & 1/3 & 0 \\ 0 & -4/3 & 1 \\ 0 & 1 & 2 \end{pmatrix} \begin{pmatrix} x_1 \\ x_2 \\ x_3 \end{pmatrix} = \begin{pmatrix} 1/3 \\ 2/3 \\ 1 \end{pmatrix}.$$

Since $a_{31} = 0$, we are done with column 1. The pivot in column 2 is $a_{22} = -4/3$, so divide row 2 by $-4/3$ to get

$$\begin{pmatrix} 1 & 1/3 & 0 \\ 0 & 1 & -3/4 \\ 0 & 1 & 2 \end{pmatrix} \begin{pmatrix} x_1 \\ x_2 \\ x_3 \end{pmatrix} = \begin{pmatrix} 1/3 \\ -1/2 \\ 1 \end{pmatrix}.$$

Multiply row 2 by 1/3 and subtract it from row 1 to zero out a_{12}, and subtract row 2 from row 3 to zero out a_{32}, yielding

$$\begin{pmatrix} 1 & 0 & 1/4 \\ 0 & 1 & -3/4 \\ 0 & 0 & 11/4 \end{pmatrix} \begin{pmatrix} x_1 \\ x_2 \\ x_3 \end{pmatrix} = \begin{pmatrix} 1/2 \\ -1/2 \\ 3/2 \end{pmatrix}.$$

We are done with column 2. The pivot in column 3 is $a_{33} = 11/4$, so divide row 3 by 11/4 to get

$$\begin{pmatrix} 1 & 0 & 1/4 \\ 0 & 1 & -3/4 \\ 0 & 0 & 1 \end{pmatrix} \begin{pmatrix} x_1 \\ x_2 \\ x_3 \end{pmatrix} = \begin{pmatrix} 1/2 \\ -1/2 \\ 6/11 \end{pmatrix}.$$

Zeroing out a_{13} and a_{23} produces the system

$$\begin{pmatrix} 1 & 0 & 0 \\ 0 & 1 & 0 \\ 0 & 0 & 1 \end{pmatrix} \begin{pmatrix} x_1 \\ x_2 \\ x_3 \end{pmatrix} = \begin{pmatrix} 4/11 \\ -1/11 \\ 6/11 \end{pmatrix},$$

and hence

$$\mathbf{x} = \begin{pmatrix} 4/11 \\ -1/11 \\ 6/11 \end{pmatrix}.$$

Gauss–Jordan elimination is the most convenient method to find the inverse $\mathbf{A}^{-1}$ of a matrix $\mathbf{A}$. Consider, for example, the 3×3 case, where we are given a matrix $\mathbf{A}$, and thus the 9 entries of $\mathbf{A}^{-1}$ are the unknowns.

$$\mathbf{A}\mathbf{A}^{-1} = \mathbf{I} \implies \begin{pmatrix} a_{11} & a_{12} & a_{13} \\ a_{21} & a_{22} & a_{23} \\ a_{31} & a_{32} & a_{33} \end{pmatrix} \begin{pmatrix} \tilde{a}_{11} & \tilde{a}_{12} & \tilde{a}_{13} \\ \tilde{a}_{21} & \tilde{a}_{22} & \tilde{a}_{23} \\ \tilde{a}_{31} & \tilde{a}_{32} & \tilde{a}_{33} \end{pmatrix} = \begin{pmatrix} 1 & 0 & 0 \\ 0 & 1 & 0 \\ 0 & 0 & 1 \end{pmatrix}.$$

This is equivalent to solving the following three systems of equations:

$$\mathbf{A} \cdot \begin{pmatrix} \tilde{a}_{11} \\ \tilde{a}_{21} \\ \tilde{a}_{31} \end{pmatrix} = \begin{pmatrix} 1 \\ 0 \\ 0 \end{pmatrix}, \quad \mathbf{A} \cdot \begin{pmatrix} \tilde{a}_{12} \\ \tilde{a}_{22} \\ \tilde{a}_{32} \end{pmatrix} = \begin{pmatrix} 0 \\ 1 \\ 0 \end{pmatrix}, \quad \text{and } \mathbf{A} \cdot \begin{pmatrix} \tilde{a}_{13} \\ \tilde{a}_{23} \\ \tilde{a}_{33} \end{pmatrix} = \begin{pmatrix} 0 \\ 0 \\ 1 \end{pmatrix}.$$

Gauss–Jordan elimination makes it easy to do all of these solves at once.

Example 36 (Gauss-Jordan elimination to find an inverse). *Use Gauss–Jordan elimination to find the inverse of the matrix*

$$A = \begin{pmatrix} 1 & 1 & 2 \\ 3 & 4 & 1 \\ 1 & 2 & 0 \end{pmatrix}.$$

We want to solve the systems of equations given by

$$\begin{pmatrix} 1 & 1 & 2 \\ 3 & 4 & 1 \\ 1 & 2 & 0 \end{pmatrix} A^{-1} = \begin{pmatrix} 1 & 0 & 0 \\ 0 & 1 & 0 \\ 0 & 0 & 1 \end{pmatrix}.$$

Use Gauss-Jordan elimination to reduce A to the identity matrix I. Zero out the off-diagonal entries in column 1 to get

$$\left(\begin{array}{ccc|ccc} 1 & 1 & 2 & 1 & 0 & 0 \\ 0 & 1 & -5 & -3 & 1 & 0 \\ 0 & 1 & -2 & -1 & 0 & 1 \end{array} \right).$$

Zero out the off-diagonal entries in column 2 to get

$$\left(\begin{array}{ccc|ccc} 1 & 0 & 7 & 4 & -1 & 0 \\ 0 & 1 & -5 & -3 & 1 & 0 \\ 0 & 0 & 3 & 2 & -1 & 1 \end{array} \right).$$

The pivot in column 3 is equal to 3, so we first divide row 3 by 3 to normalize:

$$\left(\begin{array}{ccc|ccc} 1 & 0 & 7 & 4 & -1 & 0 \\ 0 & 1 & -5 & -3 & 1 & 0 \\ 0 & 0 & 1 & 2/3 & -1/3 & 1/3 \end{array} \right).$$

Finally, zero out the off-diagonal entries in column 3 to get

$$\left(\begin{array}{ccc|ccc} 1 & 0 & 0 & -2/3 & 4/3 & -7/3 \\ 0 & 1 & 0 & 1/3 & -2/3 & 5/3 \\ 0 & 0 & 1 & 2/3 & -1/3 & 1/3 \end{array} \right).$$

A^{-1} *is then the matrix on the RHS of the last equation.*

If you know A^{-1} for some matrix A, you can solve $Ax = b$ by solving the equivalent system $A^{-1}Ax = A^{-1}b$, which simplifies to $x = A^{-1}b$. However, **using a matrix inverse to solve a linear system is a bad idea if the matrices are not tiny!** The reason for this is that most large linear system matrices are sparse, and regular Gauss elimination (and many other methods not discussed herein) can take advantage of this. The worst problem with using inverses is storage. Even if a matrix A is sparse and banded, its inverse is typically full, and even storing a 10,000 × 10,000 matrix on your computer may not be possible. Since systems of modern interest are bigger than million × million, it is easy to see that using inverses is not a viable option.

The syntax to find the inverse of A in MATLAB is `inv(A)`.

6.2 Matrix and vector norms and condition number

There are two major sources of error that arise when solving linear systems of equations $\mathbf{Ax} = \mathbf{b}$. The first comes from poor representation of the equations in the computer. This arises in the 16th digit from rounding error, but also if the equations are created from experiments, then likely there is measurement error in the fourth (or so) digit in each entry of $\mathbf{A}$ and $\mathbf{b}$. Hence although one wants to solve $\mathbf{Ax} = \mathbf{b}$, one is really solving $\hat{\mathbf{A}}\hat{\mathbf{x}} = \hat{\mathbf{b}}$. The question then arises, how close is $\hat{\mathbf{x}}$ to $\mathbf{x}$?

The second source of error comes from the calculations that produce a numerical solution to $\mathbf{Ax} = \mathbf{b}$. When Gaussian elimination (or some variant of it) is used as the linear solver, the numerical error produced is generally in the last few digits of a number (i.e. relative error is small). With other types of solvers such as Conjugate Gradient method (which we do not discuss in this book), there are residual tolerances and the error can typically occur in the sixth significant digit or worse. Since we do not discuss such solvers in this book, we do not consider this error source herein, but we emphasize that if such solvers are used then it absolutely needs to be considered.

Thus, we focus on solving $\mathbf{Ax} = \mathbf{b}$ when there is some error in the entries of $\mathbf{A}$ or $\mathbf{b}$, and we assume the numerical error from the solving process is negligible. This section discusses some basic ideas about how to quantify this error, which will let us know whether we can expect the answers we get to be any good.

In order to study how 'close' two vectors are, we will need to introduce norms in order to measure the distance between two vectors and get back a number instead of a vector. Some commonly used vector norms are:

1. Infinity norm: $\|\mathbf{v}\|_\infty = \max\limits_{1 \le i \le n} |v_i|$;

2. Euclidean norm (2-norm): $\|\mathbf{v}\|_2 = \left(\sum_{i=1}^n v_i^2 \right)^{1/2}$.

Although the intuitive measure of distance is the 2-norm, mathematically speaking, any function that satisfies the following four properties can be considered a reasonable measure of distance:

1. $\|x\| \ge 0$;
2. $\|x\| = 0$ if and only if $x = 0$;
3. $\|\alpha x\| = |\alpha| \, \|x\|$ for all $\alpha \in \mathbb{R}$;
4. $\|x + y\| \le \|x\| + \|y\|$.

The reason why we define a different way to measure the 'size' of a vector is that all measures in finite dimension are equivalent (not equal, however), and so we will want to pick one that is easy to calculate.

We will also need to discuss the 'size' of a matrix. We will consider only matrix norms induced by vector norms, i.e.

$$\|\mathbf{A}\|_i = \max_{\|\mathbf{x}\|_i = 1} \|\mathbf{Ax}\|_i,$$

where i could be 2, or ∞. Under this definition, induced matrix norms satisfy
1. $\|A\| = 0$ if and only if $A = 0$;
2. $\|\alpha A\| = |\alpha| \, \|A\|$ for all $\alpha \in \mathbb{R}$;
3. $\|Ax\| \le \|A\| \, \|x\|$;
4. $\|A + B\| \le \|A\| + \|B\|$.

Some more commonly used induced matrix norms are:
1. infinity norm (max absolute row sum): $\|A\|_\infty = \max\limits_{1 \le i \le n} \sum_{j=1}^{n} |a_{ij}|$;
2. 2-norm: $\|A\|_2 = \max\limits_{x \in \mathbb{R}^n} \dfrac{\|Ax\|_2}{\|x\|_2}$.

There is no simple formula currently known for the 2 norm. For this reason, taking matrix norms is usually done with the infinity norm.

Example 37. *Find $\|A\|_\infty$ and $\|v\|_\infty$ for*

$$A = \begin{pmatrix} 1 & -3 & 7 \\ -3 & 1 & 10 \\ -10 & 8 & -4 \end{pmatrix}, \quad v = \begin{pmatrix} 1 \\ -6 \\ 5 \end{pmatrix}.$$

For the vector norm, we calculate

$$\|v\|_\infty = \max\{1,\ 6,\ 5\} = 6,$$

For the matrix norm, again we calculate

$$\|A\|_\infty = \max\{11,\ 14,\ 22\} = 22,$$

We can now define the condition number of a matrix:

$$\operatorname{cond}(A) = \|A\| \, \left\|A^{-1}\right\|.$$

Clearly $\operatorname{cond}(A)$ depends on the choice of norm. However, due to equivalence of norms, no matter which one you pick, you will generally get $\operatorname{cond}(A)$ to be of the same order of magnitude, which is sufficient for our use of it. Hence for simplicity, we will consider only the infinity norm.

The condition number satisfies $1 \le \operatorname{cond}(A) \le \infty$, and its value increases as the rows of A get closer to being linear dependent. That is, if we think of the rows of a matrix as n dimensional vectors, if they are all perpendicular to each other, the condition number is 1. However, as the smallest angle made by the vectors shrinks, the condition number grows inversely proportional to the smallest angle.

As we will see in the next section, the condition number is of fundamental importance in linear system solving. In short, it is a measure of sensitivity in solving linear systems:

If the condition number of **A** is small:
- Numerical solution should be good:
- **Ax** = **b** is called a **well-conditioned** system.

If the condition number of **A** is large:
- Numerical solution likely bad:
- **Ax** = **b** is called an **ill-conditioned** system.

6.3 Sensitivity in linear system solving

We begin this section with an example.

Example 38. *Consider the system of equations*

$$\mathbf{Ax} = \begin{pmatrix} 6 & -2 \\ 11.5 & -3.85 \end{pmatrix} \begin{pmatrix} x_1 \\ x_2 \end{pmatrix} = \begin{pmatrix} 10 \\ 17 \end{pmatrix} = \mathbf{b}.$$

The solution to this system is

$$\mathbf{x} = \begin{pmatrix} 45 \\ 130 \end{pmatrix}.$$

However, for the slightly perturbed system

$$\widetilde{\mathbf{A}}\mathbf{x} = \begin{pmatrix} 6 & -2 \\ 11.5 & -3.84 \end{pmatrix} \begin{pmatrix} x_1 \\ x_2 \end{pmatrix} = \begin{pmatrix} 10 \\ 17 \end{pmatrix} = \mathbf{b},$$

the solution to the system is

$$\mathbf{x} = \begin{pmatrix} 110 \\ 325 \end{pmatrix}.$$

Why is there such a difference in solutions? After all, we made just a 'small' change to one entry of *A*. The reason comes down to the sensitivity of a matrix, which can be measured by its condition number. If a matrix is ill-conditioned (large condition number), then small changes to data can cause large changes in solutions. Obviously, we want to be aware of such problems. This fact can be mathematically described as follows.

Theorem 39. *Let* **Ax** = **b** *and* **Âx̂** = **b**. *Then*

$$\frac{\|\mathbf{x} - \hat{\mathbf{x}}\|}{\|\hat{\mathbf{x}}\|} \leq \mathrm{cond}(\mathbf{A}) \frac{\|\mathbf{A} - \hat{\mathbf{A}}\|}{\|\mathbf{A}\|}.$$

Proof. Since **b** = **Ax** and **b** = **Âx̂**, **Ax** = **Âx̂**, and subtracting **Ax̂** from both sides gives

$$\mathbf{A}(\mathbf{x} - \hat{\mathbf{x}}) = (\hat{\mathbf{A}} - \mathbf{A})\hat{\mathbf{x}}.$$

Multiplying both sides by A^{-1} and taking norms of both sides gives

$$\|x - \hat{x}\| = \|A^{-1}\|\|(\hat{A} - A)\hat{x}\| \le \|A^{-1}\|\|\hat{A} - A\|\|\hat{x}\|,$$

with the inequality coming from the matrix norm properties. Next, divide both sides by $\|\hat{x}\|$, and multiply the right hand side by $1 = \frac{\|A\|}{\|A\|}$ to reveal

$$\frac{\|x - \hat{x}\|}{\|\hat{x}\|} \le \|A^{-1}\|\|\hat{A} - A\|\frac{\|A\|}{\|A\|}.$$

Inserting the definition of condition number of A, we get

$$\frac{\|x - \hat{x}\|}{\|\hat{x}\|} \le \text{cond}(A)\frac{\|\hat{A} - A\|}{\|A\|},$$

which is the stated result. $\qquad\square$

The consequence of this theorem is that the relative change made to A is magnified in the solution by $\text{cond}(A)$. In fact, you can even think of the theorem as saying the following: the percent difference between the solutions is bounded by the percent difference between the matrices times the condition number.

Note there is a similar result for changes in the vector b, which can be proved and interpreted in a similar manner.

Theorem 40. *Let* $Ax = b$ *and* $A\hat{x} = \hat{b}$. *Then*

$$\frac{\|x - \hat{x}\|}{\|x\|} \le \text{cond}(A)\frac{\|b - \hat{b}\|}{\|b\|}.$$

In general, for a given linear system of equations, the best estimate we can hope for is

Relative error in numerical solution $\approx \text{cond}(A) \cdot$ machine epsilon.

- This ignores possible error in numerical methods for $\hat{x}$, so error could be worse.
- If **b** comes from data measurements, then

Relative error in numerical solution $\approx \text{cond}(A) \cdot$ data measurement error.

Thus, knowing $\text{cond}(A)$ is very important.

Returning now to our example, we can answer the question of why the two solution differed by so much. We calculate the condition number first, via

$$A^{-1} = \begin{pmatrix} 38.5 & -20 \\ 115 & -60 \end{pmatrix},$$

$$\text{cond}(\mathbf{A}) = \|\mathbf{A}\|_\infty \|\mathbf{A}^{-1}\|_\infty$$

$$= \max\{8, 15.35\} \cdot \max\{58.5, 175\}$$

$$= 15.35 \cdot 175$$

$$= 2686.25.$$

The relative change in A is $\frac{\|A - \bar{A}\|}{\|A\|} \approx \frac{0.01}{15.35} = 6.51e - 4$. Hence from the theorem, we can expect a relative difference between $\mathbf{x}$ and $\hat{\mathbf{x}}$ to be bounded by

$$\frac{\|\mathbf{x} - \hat{\mathbf{x}}\|}{\|\hat{\mathbf{x}}\|} \leq 2.69e + 3 \cdot 6.51e - 4 = 1.752,$$

and thus we can only expect the relative change to be less than 175%, which is about what we observe. Of course, this is terrible, and is caused directly by the condition number being large relative to the size of the perturbation.

6.4 Exercises

1. Given $\mathbf{A} = \begin{pmatrix} 2 & 4 \\ -3 & -6.001 \end{pmatrix}$ and $\mathbf{b} = \begin{pmatrix} 2 \\ 3 \end{pmatrix}$:
   (a) Solve $Ax = b$ using backslash.
   (b) Change the second entry of b to 3.01, and solve for the new solution.
   (c) What is the relative difference between the solutions?
   (d) Does this agree with the theorem about the difference between solutions when the right hand side vector has been perturbed?

2. Repeat exercise 1, but now using the matrix $\mathbf{A} = \begin{pmatrix} 2 & 4 \\ 3 & 6.001 \end{pmatrix}$

3. Use Gauss-Jordan elimination by hand to find the inverse of

$$A = \begin{pmatrix} 1 & 2 & 3 \\ 4 & 1 & 2 \\ 0 & 1 & 5 \end{pmatrix}.$$

4. Prove Theorem 40.

7 Eigenvalues and eigenvectors

Eigenvalues and eigenvectors are important in many applications. The natural modes and frequencies of a mechanical structure can be determined by eigenvalues and eigenvectors of certain matrices. The reproduction rate of an infectious disease in a population is determined by the largest eigenvalue of a matrix. The reaction rate of a chemical or nuclear system is often determined by the largest eigenvalue of a matrix. Moreover, many application problems are naturally formulated as eigenvalue problems.

One common use of eigenvalues is to find all the roots of a polynomial. If

$$p(x) = c_0 + c_1 x + c_2 x^2 + \cdots + c_{n-1} x^{n-1} + x^n,$$

then the eigenvalues of the **companion matrix**

$$\begin{pmatrix} 0 & 0 & \cdots & \cdots & 0 & -c_0 \\ 1 & 0 & 0 & \cdots & 0 & -c_1 \\ 0 & 1 & 0 & \cdots & 0 & -c_2 \\ \vdots & \vdots & \vdots & \vdots & \vdots & \vdots \\ 0 & \cdots & \cdots & 0 & 1 & -c_{n-1} \end{pmatrix}$$

are the roots of p. Note that since formulas like quadratic, cubic, and quartic are prone to catastrophic round-off error, finding roots this way is usually the most reliable approach. There is no formula for higher order polynomial roots, so in this sense finding the roots as eigenvalues circumvents this problem.

7.1 Mathematical definition

An eigenvalue-eigenvector pair for a matrix A is a scalar λ and vector $x \neq 0$ satisfying

$$Ax = \lambda x,$$

Even if A is real, the λ and x may be complex. Also, note that eigenvectors are not unique (e.g. if x is an eigenvector, then so is $-x$ and $2x$). Generally, eigenvectors are normalized to have length one and that their first nonzero component be positive.

An important part of the definition (that is easy to forget) is that x must not be the zero vector. The zero vector would make every λ fullfill the equation $Ax = \lambda x$. Note that zero eigenvalues are perfectly fine (and appear if A is not invertible).

If we have that $Ax = \lambda x$, then

$$Ax = \lambda x \iff (A - \lambda I)x = 0.$$

This homogeneous system of equations has a nonzero solution $\mathbf{x}$ if and only if $\mathbf{A} - \lambda\mathbf{I}$ is singular. Hence, the eigenvalues of $\mathbf{A}$ are the values of λ such that

$$\det(\mathbf{A} - \lambda\mathbf{I}) = 0.$$

If $\mathbf{A}$ is $n \times n$, then $\det(\mathbf{A} - \lambda\mathbf{I})$ is a polynomial of degree n in λ, known as the **characteristic polynomial** of $\mathbf{A}$, and its roots are the eigenvalues of $\mathbf{A}$. Note that the roots of the characteristic polynomial need not be real numbers even if $\mathbf{A}$ has real entries. The fundamental theorem of algebra states that the characteristic polynomial will have exactly n roots (including complex and multiple roots), so the matrix $\mathbf{A}$ has n eigenvalues when including multiplicity.

Example 41 (Eigenvalues). *Determine the eigenvalues of the matrix*

$$\mathbf{A} = \begin{pmatrix} 2 & 1 \\ 1 & 2 \end{pmatrix}.$$

The characteristic polynomial of $\mathbf{A}$ *is*

$$0 = \det(\mathbf{A} - \lambda\mathbf{I}) = \det\left(\begin{pmatrix} 2-\lambda & 1 \\ 1 & 2-\lambda \end{pmatrix}\right) = (2-\lambda)^2 - 1 = \lambda^2 - 4\lambda + 3.$$

Factoring this equation gives

$$(\lambda - 3)(\lambda - 1) = 0,$$

and thus the eigenvalues of $\mathbf{A}$ *are* $\lambda = 3$ *and* $\lambda = 1$.

Once an eigenvalue λ is known, its corresponding eigenvector can be found by solving $(\mathbf{A} - \lambda\mathbf{I})\mathbf{x} = 0$ for a given λ. The system will not be uniquely solvable but the set of solutions will form a subspace with dimension of at least 1 and at most that of the (algebraic) multiplicity of the eigenvalue.

Example 42 (Eigenvectors). *We want to determine an eigenvector corresponding to the eigenvalue* $\lambda = 3$, *of the matrix* $\mathbf{A}$ *from our last example. Note that solving*

$$\mathbf{Ax} = \begin{pmatrix} 2 & 1 \\ 1 & 2 \end{pmatrix}\begin{pmatrix} x_1 \\ x_2 \end{pmatrix} = \begin{pmatrix} 3x_1 \\ 3x_2 \end{pmatrix} = 3\mathbf{x}$$

is equivalent to solving the system of equations

$$2x_1 + x_2 = 3x_1$$
$$x_1 + 2x_2 = 3x_2,$$

which says that $x_1 = x_2$. *Eigenvectors are not unique, so for any* $\alpha \in \mathbb{R}$, *the vector*

$$\mathbf{x} = \begin{pmatrix} \alpha \\ \alpha \end{pmatrix}$$

is an eigenvector corresponding to $\lambda = 3$. You might pick $\mathbf{x} = \begin{pmatrix} 1 \\ 1 \end{pmatrix}$ as your answer.

A similar calculation for $\lambda = 1$ will give the eigenvector $\mathbf{x} = \begin{pmatrix} 1 \\ -1 \end{pmatrix}$.

In MATLAB the eigenvalues (and eigenvectors) of a matrix $\mathbf{A}$ can be calculated using the command: `[V,D]=eig(A);`. Here the columns of the matrix V are the eigenvectors that correspond to the eigenvalues returned in D. This command is effective for matrices up to size $n = 5{,}000$ or so. Beyond that, generally speaking, the calculation becomes slow and is prone to roundoff error.

7.2 Power method

For large matrices, finding all eigenvalues and eigenvectors of a matrix is hard and expensive. Currently, $n \approx 5{,}000$ is about the limit for most matrices for which one can expect reasonable error and runtime.

However, finding just the largest eigenvalue of a matrix is usually possible to get efficiently and accurately, and often, knowing the largest eigenvalue is good enough. The **power method** is an iterative procedure for finding the largest eigenvalue and a corresponding eigenvector of a matrix.

Assume $\mathbf{A}$ has eigenvalues satisfying $|\lambda_1| > |\lambda_2| \geq |\lambda_3| \geq |\lambda_4| \cdots \geq |\lambda_n|$ (in other words, it has a single largest eigenvalue in magnitude) and linearly independent eigenvectors $\mathbf{y}_1, \mathbf{y}_2, \ldots, \mathbf{y}_n$ (usually a good assumption). Then any $\mathbf{x} \in \mathbb{R}^n$ can be expressed as a linear combination of eigenvectors of $\mathbf{A}$:

$$\mathbf{x} = \alpha_1 \mathbf{y}_1 + \alpha_2 \mathbf{y}_2 + \cdots + \alpha_n \mathbf{y}_n.$$

Multiplying both sides by $\mathbf{A}$ gives

$$\mathbf{A}\mathbf{x} = \alpha_1 \mathbf{A}\mathbf{y}_1 + \alpha_2 \mathbf{A}\mathbf{y}_2 + \cdots + \alpha_n \mathbf{A}\mathbf{y}_n = \alpha_1 \lambda_1 \mathbf{y}_1 + \alpha_2 \lambda_2 \mathbf{y}_2 + \cdots \alpha_n \lambda_n \mathbf{y}_n.$$

Multiplying by $\mathbf{A}$ again gives

$$\mathbf{A}^2 \mathbf{x} = \alpha_1 \lambda_1^2 \mathbf{y}_1 + \alpha_2 \lambda_2^2 \mathbf{y}_2 + \cdots \alpha_n \lambda_n^2 \mathbf{y}_n.$$

Repeating this m times gives

$$\mathbf{A}^m \mathbf{x} = \alpha_1 \lambda_1^m \mathbf{y}_1 + \alpha_2 \lambda_2^m \mathbf{y}_2 + \cdots \alpha_n \lambda_n^m \mathbf{y}_n.$$

Since $|\lambda_1| > |\lambda_2| \geq |\lambda_3| \geq \cdots \geq |\lambda_n|$, for m large enough we have

$$\mathbf{A}^m \mathbf{x} \approx \alpha_1 \lambda_1^m \mathbf{y}_1.$$

Using this approximation we get

$$\mathbf{A}\mathbf{A}^m \mathbf{x} = \mathbf{A}^{m+1} \mathbf{x} = \alpha_1 \lambda_1^{m+1} \mathbf{y}_1 = \lambda_1 \alpha_1 \lambda_1^m \mathbf{y}_1 = \lambda_1 \mathbf{A}^m \mathbf{x},$$

which means $\mathbf{A}^m x$ is an eigenvector of $\mathbf{A}$. The power method now involves repeatedly applying $\mathbf{A}$ to the starting vector x and normalizing it in every step:

$$\mathbf{x}_{k+1} = \frac{\mathbf{A}\mathbf{x}_k}{\|\mathbf{A}\mathbf{x}_k\|}.$$

If the method converges (meaning $\mathbf{x}_{k+1}$ is nearly the same as $\mathbf{x}_k$) we have converged to the direction of the eigenvector. Thus we now know the $\mathbf{x}$ from $\mathbf{A}\mathbf{x} = \lambda_1\mathbf{x}$. To recover λ_1, dot product both sides of this equation with $\mathbf{x}$ to get see that

$$\lambda_1 = \frac{\mathbf{x} \cdot (\mathbf{A}\mathbf{x})}{\mathbf{x} \cdot \mathbf{x}}.$$

Algorithm 43 (Normalized power method). *Choose a nonzero vector* $\mathbf{x}_0$.

for $k = 0, 1, 2, \ldots$
  $\mathbf{y}_{k+1} = \mathbf{A}\mathbf{x}_k$
  $\mathbf{x}_{k+1} = \frac{\mathbf{y}_{k+1}}{\|\mathbf{y}_{k+1}\|}$
end
return eigenvector $\mathbf{x}_{k+1}$ *and eigenvalue* $\frac{\mathbf{x}_k^T \mathbf{A}\mathbf{x}_k}{\mathbf{x}_k^T \mathbf{x}_k}$.

It does not matter which norm you choose to use in Algorithm 43, but you should be consistent with the choice as different norms will scale the vector differently.

Example 44 (Power method). *Define $\mathbf{A}$ as in the previous two examples and let*

$$\mathbf{x}_0 = \begin{pmatrix} 1 \\ 0 \end{pmatrix}.$$

One iteration of the power method (using the infinity norm) gives

$$\mathbf{y}_1 = \mathbf{A}\mathbf{x}_0 = \begin{pmatrix} 2 & 1 \\ 1 & 2 \end{pmatrix}\begin{pmatrix} 1 \\ 0 \end{pmatrix} = \begin{pmatrix} 2 \\ 1 \end{pmatrix} \implies \mathbf{x}_1 = \begin{pmatrix} 1 \\ 1/2 \end{pmatrix}.$$

Another power iteration gives

$$\mathbf{y}_2 = \mathbf{A}\mathbf{x}_1 = \begin{pmatrix} 2 & 1 \\ 1 & 2 \end{pmatrix}\begin{pmatrix} 1 \\ 1/2 \end{pmatrix} = \begin{pmatrix} 5/2 \\ 2 \end{pmatrix} \implies \mathbf{x}_2 = \begin{pmatrix} 1 \\ 4/5 \end{pmatrix}.$$

Repeating this process 10 times (i.e. after 10 power iterations) we see that

$$\mathbf{x}_{10} = \begin{pmatrix} 1 \\ 0.999966 \end{pmatrix}.$$

Hence we see that $\mathbf{x}_n$ is converging to

$$\mathbf{x} = \begin{pmatrix} 1 \\ 1 \end{pmatrix},$$

which we know is an eigenvector of **A** *from the last example. We know that* **x** *corresponds to the eigenvalue* $\lambda = 3$, *but if we didn't we could easily figure it out by calculating*

$$\mathbf{A}\mathbf{x}_{10} = \begin{pmatrix} 2.9996613 \\ 2.9993226 \end{pmatrix},$$

which we see is approximately $3\mathbf{x}_{10}$.

Monitoring convergence of the power method on a computer can be tricky if we only use the eigenvector approximation $\mathbf{x}_k$. The issue can be that if λ_1 is negative (or has negative real part), then the sign of x_k will alternate at each step, even if $\mathbf{x}_k$ has converged to the correct direction (recall eigenvectors are not unique and if $\mathbf{x}$ is an eigenvector then so is $-\mathbf{x}$). Thus even though convergence to an eigenvector has occurred, the algorithm as stated above will not know this. Hence, it is better to watch the eigenvalue approximation $\lambda_k = \frac{\mathbf{x}_k^T \mathbf{A} \mathbf{x}_k}{\mathbf{x}_k^T \mathbf{x}_k}$.

A code for the power method is given below.

```
function [x,l] = PowerMethod(A, x0, tolerance)

x = x0;

l = x' * (A*x) / (x'*x);
lold = l+1;

itcount=0;

while abs(l - lold)>tolerance

    itcount=itcount+1;
    if itcount>100
        error('iterations exceeded 100');
    end

    y = A*x;
    x = y / norm(y,inf)

    lold = l;
    l = x' * (A*x) / (x'*x);

end
```

Convergence of the power method is related to the difference $|\lambda_1| - |\lambda_2|$. If $|\lambda_1|$ is much larger than $|\lambda_2|$ (i.e. $|\lambda_1| - |\lambda_2| \gg 0$) then the power method converges to an eigenvector of **A** in only a few iterations. But if $|\lambda_1|$ is not much larger than $|\lambda_2|$ (i.e. $|\lambda_1| - |\lambda_2| \approx 0$) then the power method will take many iterations to converge (if at all).

Sometimes the smallest eigenvalue of a matrix is needed instead of the largest. Recall any eigenvalue/eigenvector pair $(\lambda, \mathbf{x})$ satisfies $\mathbf{A}\mathbf{x} = \lambda \mathbf{x}$. Multiplying by $\mathbf{A}^{-1}$

gives $\mathbf{x} = \lambda \mathbf{A}^{-1}\mathbf{x}$, or equivalently, $\mathbf{A}^{-1}\mathbf{x} = \frac{1}{\lambda}\mathbf{x}$. For any eigenvalue λ of $\mathbf{A}$, $1/\lambda$ is an eigenvalue of $\mathbf{A}^{-1}$. Thus, finding the largest eigenvalue of $\mathbf{A}^{-1}$ will give you the smallest eigenvalue of $\mathbf{A}$. To find the smallest eigenvalue of $\mathbf{A}$, simply use the power method with $\mathbf{A}^{-1}$. But in the step where we would use the inverse,

$$\mathbf{y}_k = \mathbf{A}^{-1}\mathbf{x}_{k-1},$$

we would instead find y by solving the linear system $\mathbf{A}\mathbf{y}_k = \mathbf{x}_{k-1}$ (recall forming inverses is bad!).

Once the maximum eigenvalue λ_1 and its eigenvector $\mathbf{x}_1$ have been found, the system of equations can be adjusted so that λ_1 is no longer an eigenvalue. We can then use the power method on the new system to find the next largest eigenvalue λ_2. In this way, we can easily find the few biggest eigenvalues/eigenvectors of $\mathbf{A}$.

7.3 Application: Population dynamics

In population systems with approximately constant birth and death rates, we can develop matrices that show how the population changes from year to year. As we will show, the largest eigenvalue of this system determines the stability of the system.

We consider the following example. Suppose we have a population of animals, and each individual is classified as being in group P_1 if less than 1 year old, P_2 if between 1 and 2, P_3 if between 2 and 3, and P_4 if older than 3. The death rates, i.e. the chance that an individual will die in the next year, for each of these groups are d_1, d_2. d_3 and d_4, respectively. The birth rates for each group are b_1, b_2. b_3 and b_4. Thus if we know the size of the groups at year n, then we will know the size of the groups at year $n + 1$ from the equations

$$P_1^{n+1} = b_1 P_1^n + b_2 P_2^n + b_3 P_3^n + b_4 P_4^n$$
$$P_2^{n+1} = (1 - d_1)P_1^n$$
$$P_3^{n+1} = (1 - d_2)P_2^n$$
$$P_4^{n+1} = (1 - d_3)P_3^n + (1 - d_4)P_4^n.$$

Writing this as matrices and vectors,

$$\begin{pmatrix} P_1^{n+1} \\ P_2^{n+1} \\ P_3^{n+1} \\ P_4^{n+1} \end{pmatrix} = \begin{pmatrix} b_1 & b_2 & b_3 & b_4 \\ (1-d_1) & 0 & 0 & 0 \\ 0 & (1-d_2) & 0 & 0 \\ 0 & 0 & (1-d_3) & (1-d_4) \end{pmatrix} \begin{pmatrix} P_1^n \\ P_2^n \\ P_3^n \\ P_4^n \end{pmatrix}.$$

Using this relationship, we can determine the long time behavior of the population with repeated multiplication, since the year n population is related to year 0 population by

$$
\begin{pmatrix} P_1^n \\ P_2^n \\ P_3^n \\ P_4^n \end{pmatrix} = \begin{pmatrix} b_1 & b_2 & b_3 & b_4 \\ (1-d_1) & 0 & 0 & 0 \\ 0 & (1-d_2) & 0 & 0 \\ 0 & 0 & (1-d_3) & (1-d_4) \end{pmatrix}^n \begin{pmatrix} P_1^0 \\ P_2^0 \\ P_3^0 \\ P_4^0 \end{pmatrix}.
$$

Due to the repeated multiplication is by the same matrix, this process is exactly the (non-normalized) power method! Thus if the biggest eigenvalue is bigger than 1, then the population will grow to infinity as $n \to \infty$, and if the biggest eigenvalue is less than 1, then the population will eventually become extinct. A stable population will have a biggest eigenvalue of 1.

7.4 Exercises

1. Use the `eig` function power method to find the largest eigenvalue and corresponding eigenvector for

    ```
    A = [1 2 3; 4 5 6; 7 8 8]
    ```

    You can choose the starting vector. Use the `eig` function (i.e. 'eig(A)') to verify that your answer is correct.
2. Repeat exercise 1 for the matrix

    ```
    B = [2    3    2;  1    0    -2;  -1    -3    -1]
    ```

    using the initial guess [2;3;2]. What happens and why? (hint: use 'eig' again)
3. With the same **B** from part 3, try running the power method method with the starting vector $\mathbf{x} = [1; -1; 1]$. What happened? Can you explain why?
4. Repeat 1. and 2., but now find the smallest eigenvalue of **A** by finding the largest eigenvalue of A^{-1}. Do not compute A^{-1}, instead use linear solves. Note that only a single character needs changed in the power method code above in order to do this problem.
5. Let $a = 1e - 8$, $b = 100$, and $c = 1e - 2$. Find the roots of the quadratic $f(x) = ax^2 + bx + c$
    (a) using MATLAB and the quadratic formula (i.e. type the formula into MATLAB);
    (b) by finding the eigenvalues of the companion matrix (see the beginning of this section for the definition of the companion matrix);
    (c) which is more accurate?
6. Consider the biology example at the end of this chapter. Suppose we have birth rates $b_1 = .3$, $b_2 = .3$, $b_3 = .3$, $b_4 = .1$ and death rates $d_1 = .1$, $d_2 = .2$, $d_3 = .5$, and $d_4 = .9$.

(a) What is the biggest eigenvalue of the resulting matrix?

(b) Based on this, what do you expect to happen to the population size over a long time period?

(c) Suppose at year 0, you have $P_1^0 = 100$, $P_2^0 = 200$, $P_3^0 = 150$, $P_4^0 = 75$. What will be the population distribution at year 1000? Does this answer agree with your answer to part b?

(d) Suppose we change the problem slightly so that animals in P_4 have a death rate of $d_4 = 0.01$ instead of 0.9. Does this change the long time dynamics of the system?

8 Fitting curves to data

It is both helpful and convenient to express relations in data with functions. This allows for taking derivatives, integrating, and even solving differential equations. In this chapter we will look at the two common classes of such methods: interpolating functions and curves of best fit.

8.1 Interpolation

An interpolant of a set of points is a function that passes through each of the data points. For example, given the points $(0, 0)$, $(\pi/2, 1)$, and $(-\pi/2, -1)$, both

$$f(x) = \sin(x)$$

and

$$g(x) = \frac{2x}{\pi}$$

would be interpolating functions, as shown in the plot below.

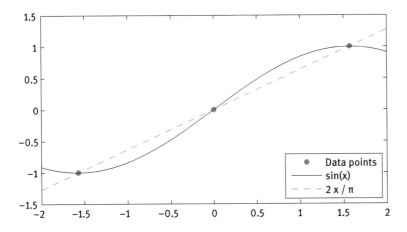

There are many instances where we would prefer to describe data with a function, since this allows us to represent the data between the points, as well as take derivatives and integrals. We will consider interpolation by a single polynomial, as well as piecewise-polynomial interpolation.

8.1.1 Interpolation by a single polynomial

Given n points: (x_i, y_i), $i = 1, 2, \ldots, n$, a degree $n - 1$ polynomial can be found that passes through each of the n points. Such a polynomial would then be an interpolant of the data points. Since we expect the interpolating polynomial to be degree $n - 1$, it must be of the form

$$p(x) = c_1 + c_2 x + c_3 x^2 + \cdots + c_n x^{n-1}.$$

Thus, there are n unknowns $(c_1, \ldots, c_n)$ that, if we could determine them, would give us the interpolating polynomial.

Requiring $p(x)$ to interpolate the data leads to n equations. That is, if p is to pass through data point i, then it must hold that $y_i = p(x_i)$. Thus, for each i ($1 \leq i \leq n$), we get the equation

$$y_i = c_1 + c_2 x_i + c_3 x_i^2 + \cdots + c_n x_i^{n-1}.$$

Since all the x_i's and y_i's are known, each of these n equations is linear in the unknown c_j's. Thus we have a square linear system that can be solved to determine the unknown coefficients:

$$\begin{pmatrix} 1 & x_1 & x_1^2 & \cdots & x_1^{n-1} \\ 1 & x_2 & x_2^2 & \cdots & x_2^{n-1} \\ 1 & x_3 & x_3^2 & \cdots & x_3^{n-1} \\ \vdots & \vdots & \vdots & \cdots & \vdots \\ 1 & x_n & x_n^2 & \cdots & x_n^{n-1} \end{pmatrix} \begin{pmatrix} c_1 \\ c_2 \\ c_3 \\ \vdots \\ c_n \end{pmatrix} = \begin{pmatrix} y_1 \\ y_2 \\ y_3 \\ \vdots \\ y_n \end{pmatrix}.$$

Solving this linear system will determine the polynomial! Consider the following example:

Example 45. *Find a polynomial that interpolates the five points*

$$(0, 1), (1, 2), (2, 2), (3, 6), (4, 9).$$

```
% x and y values of the 5 points to interpolate:
px = [0 1 2 3 4];
py = [1 2 2 6 9];
n = length(px);

% build nxn matrix column by column
A = zeros(n,n);
for i=1:n
  A(:,i) = px.^(i-1);
end

% now solve for the coefficients:
c = A\py';

% output matrix and coefficients:
A
c
```

```
% plot the points and the polynomial at the points x
x = linspace(-1,5,100);
y = c(1) + c(2)*x + c(3)*x.^2 + c(4) *x.^3 + c(5)*x.^4;
plot(px,py,'rs',x,y,'k')
```

This code will produce the output:

```
A =
     1      0      0      0      0
     1      1      1      1      1
     1      2      4      8     16
     1      3      9     27     81
     1      4     16     64    256


c =
    1.000000000000000
    5.666666666666661
   -7.583333333333332
    3.333333333333333
   -0.416666666666667
```

Thus we have found our interpolating polynomial for the 5 points:

$$p(x) = 1.000 + 5.666x - 7.583x^2 + 3.333x^3 - 0.416x^4.$$

Plotting it along with the original data points gives the plot below, from which we can see that the curve does indeed pass through each point.

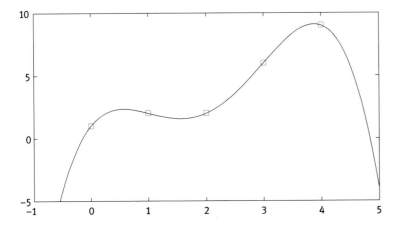

If n is larger than a few, then typically it is a bad idea to interpolate data with a single polynomial. First of all, very few processes are correctly modeled with high order polynomials. Also, higher order polynomials are inherently oscillatory, and thus even if data is from a simple function, small amounts of noise or error can cause the interpolating polynomial to oscillate between data points. Hence this type of interpolation is only useful when n is large there is no data error present. Futhermore, if $n \geq 10$ or so, the conditioning of the matrix becomes bad, and thus there may also be error in the linear solve that produces the coefficients. However, this last point can be fixed by using Lagrange or Newton interpolation, which we do not discuss herein; since we have already provided two good reasons not to use higher order single polynomial interpolation, we do not expand on these other methods.

The following picture shows the oscillations if you interpolate the 10 points with y values 0 and 1:

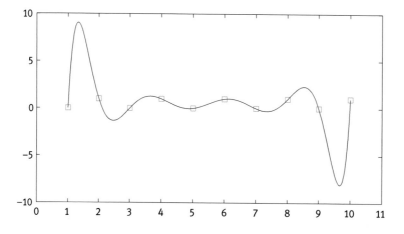

It is clear from the picture that the interpolating polynomial (probably) does not give a good description of what is happening between the data points.

8.1.2 Piecewise polynomial interpolation

As we discussed in the previous section, interpolation of many points using a single polynomial will not typically give accurate predictions due to oscillations that can occur. However, it is perfectly reasonable to instead interpolate data with many, low degree polynomials. In the picture below, we show a piecewise linear interpolation of some data points.

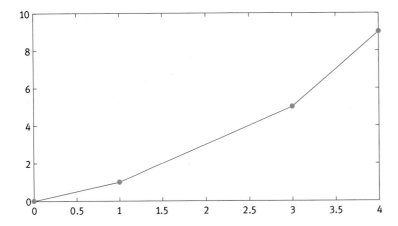

Piecewise linear interpolation is easily defined, on interval i, to be

$$p_i(x) = y_i + \frac{y_{i+1} - y_i}{x_{i+1} - x_i}(x - x_i).$$

Then the interpolating polynomial p is defined so that $p = p_i$ on interval i.

This may seem simplistic, but often it is quite accurate. It avoids oscillations, and even satisfies the following error estimate.

Theorem 46. *Suppose we pick n data points, (x_i, y_i), i=1,...,n, with max spacing h in the x-points, from a function $f \in C^2([x_1, x_n])$. Then the difference between the function and the piecewise linear interpolant satisfies*

$$\max_{x_1 \le x \le x_n} |f(x) - p(x)| \le h^2 \max_{x_1 \le x \le x_n} |f''(x)|.$$

Often a better choice of a piecewise interpolant is a cubic spline interpolant. This type of interpolant will be cubic function on each interval and will satisfy
1. on interval i,
$$p_i(x) = c_0^{(i)} + c_1^{(i)}x + c_2^{(i)}x^2 + c_3^{(i)}x^3;$$

2. $p \in C^2([x_1, x_n])$ (i.e. it is smooth, not choppy like the piecewise linear interpolant).

Determining the $4(n - 1)$ unknowns is done by enforcing that the polynomial p to interpolate the data ($2n - 2$ equations), enforcing $p \in C^2$ at nodes x_2 through x_{n-1} ($2n - 4$ equations), and enforcing $p''(x_1) = p''(x_n) = 0$ (2 equations). This provides a $(4n - 4) \times (4n - 4)$ square linear system that we can solve to determine each of the $c_j^{(i)}$'s.

If the data points come from a smooth function, then cubic splines can be much more accurate than piecewise linear interpolants.

Theorem 47. *Suppose we pick n data points, (x_i, y_i), $i = 1, \ldots, n$, with max point spacing h in the x-points, from a function $f \in C^4([x_1, x_n])$. Then the difference between the* *function and the piecewise linear interpolant satisfies*

$$\max_{x_1 \le x \le x_n} |f(x) - p(x)| \le Ch^4 \max_{x_1 \le x \le x_n} |f''''(x)|.$$

MATLAB has a built in function to do this, called 'spline'. All we have to do is give it the data points, and MATLAB takes care of the rest, as is demonstrated in the following example.

Example 48. *For the data points $(i, \sin(i))$ for $i = 0, 1, 2, \ldots, 6$, find a cubic spline. Then plot the spline along with the data points and the function $\sin(x)$.*

First, we create the data points and the spline.

```
>> x = [0:6]
x =
     0    1    2    3    4    5    6
>> y = sin(x);
>> cs = spline(x,y);
```

Next, we can evaluate points in the spline using 'ppval'. So for example, we can plug in $x = \pi$ via

```
>> ppval(cs,pi)
ans =
   -1.3146e-04
```

Thus to plot the cubic spline, we just need to give it x points to plot y values at. We do this below, and include in our plot the sin *curve as well.*

```
>> xx = linspace(0,6,1000);
>> yy = ppval(cs,xx);
>> plot(x,y,'rs',xx,yy,'r',xx,sin(xx),'b')
>> legend('data points','cubic spline','sin(x)')
```

which produces the following plot:

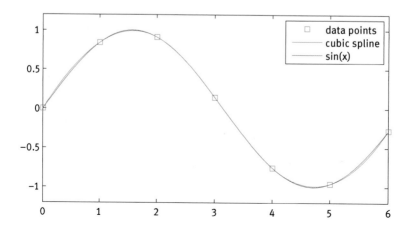

From the plot, we see that with just 7 points, we are able to match the sin *curve on a large interval quite well!*

8.2 Curve fitting

In many applications, it is not appropriate to try to have an approximating function pass through every single data point. For example, if there is a large amount of noise/error in the data, this can result in significant error. A better alternative is to fit a simple function to the data, which need not pass through any data points, but still is a good representation of the data.

We will first develop this idea for a line of best fit, and then extend to curves of best fit.

8.2.1 Line of best fit

Suppose we are given the data points (x_1, y_1), (x_2, y_2), ..., (x_n, y_n), and we wish to draw a single line that best represents the data. It is intuitive how to do this by hand (by 'eyeing it up'), but we need a mathematical framework in order to formalize a procedure.

We begin by identifying the unknowns. Since we want a line, really what we want to know is: 'What are a_0 and a_1 such that the line $y = a_0 + a_1 x$ is the line of best fit for the data?'

How to determine a_0 and a_1 depends on how we define the notion of 'best fit'. The measure of 'best' that our eyeball sees is the Euclidean norm, and so we want a_0 and a_1 so that the error

$$e(a_0, a_1) = \sum_{i=1}^{n} (y_i - (a_0 + a_1 x_i))^2$$

is minimized. Note this is the sum of squares of the differences between the data point y-value y_i, and the y-value $a_0 + a_1 x_i$ predicted by the line with coefficients a_0 and a_1.

The answer of how to minimize this function is easy. Note that geometrically the function e is simply a quadratic function (i.e. a paraboloid) in its variables a_0 and a_1:

$$e(a_0, a_1) = \sum_{i=1}^{n} (y_i - (a_0 + a_1 x_i))^2$$

$$= \sum_{i=1}^{n} y_i^2 - 2 \sum_{i=1}^{n} y_i(a_0 + a_1 x_i) + \sum_{i=1}^{n} (a_0 + a_1 x_i)^2$$

$$= \sum_{i=1}^{n} y_i^2 - 2a_0 \sum_{i=1}^{n} y_i - 2a_1 \sum_{i=1}^{n} y_i x_i + na_0^2 + a_1^2 \sum_{i=1}^{n} x_i^2 + 2a_0 a_1 \sum_{i=1}^{n} x_i .$$

Since this function is an upward facing paraboloid, we know its minimum is at its vertex, and moreover, it has only 1 local minimum which is also its global minimum.

Recall from multivariable Calculus that a local minimum of a function of 2 variables occurs when both first partial derivatives are zero. Hence to find the 'best' a_0 and a_1, we can solve the equations

$$\frac{\partial e}{\partial a_0} = 0, \quad \frac{\partial e}{\partial a_1} = 0.$$

To simplify notation, let $X = \sum_{i=1}^{n} x_i$, $Y = \sum_{i=1}^{n} y_i$, $W = \sum_{i=1}^{n} x_i^2$, and $Z = \sum_{i=1}^{n} x_i y_i$. Now we calculate

$$0 = \frac{\partial e}{\partial a_0} = -2Y + 2na_0 + 2a_1 X,$$

$$0 = \frac{\partial e}{\partial a_1} = -2Z + 2a_1 W + 2a_0 X.$$

Thus we have 2 linear equations with 2 variables, which gives us the matrix equation

$$\begin{pmatrix} n & X \\ X & W \end{pmatrix} \begin{pmatrix} a_0 \\ a_1 \end{pmatrix} = \begin{pmatrix} Y \\ Z \end{pmatrix},$$

which has a unique solution that can be determined with a linear solve.

Example 49. *Find the line of best fit for the data points*

```
xy = [
        1.0000      1.0000
        1.9000      1.0000
        2.5000      1.1000
        3.0000      1.6000
        4.0000      2.0000
        7.0000      3.4500
    ]
```

The first thing one should always do when given data is plot it. Here, we need to ask ourselves whether it is reasonable to fit a line to the data. From the plot, the answer appears to be yes.

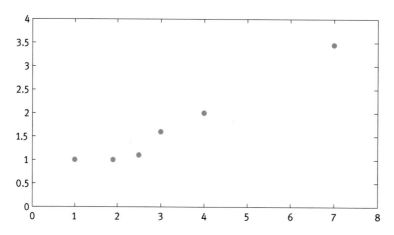

Hence we proceed to find the coefficients. First, we construct X, Y, Z, and W, using MATLAB:

```
>> X = sum(xy(:,1))
X =
   19.4000
>> Y = sum(xy(:,2))
Y =
   10.1500
>> W = sum( xy(:,1).*xy(:,1))
W =
   84.8600
>> Z = sum( xy(:,1).*xy(:,2))
Z =
   42.6000
>> n = length(xy)
n =
        6
```

Next, construct the 2 × 2 linear system, and solve it.

```
>> A = [n X;X W]
A =
    6.0000   19.4000
   19.4000   84.8600
>> b = [Y;Z]
b =
   10.1500
   42.6000
>> a = A \ b
a =
    0.2627
    0.4419
```

We now have the coefficients, $a_0 = 0.2627$ and $a_1 = 0.4419$. As a sanity check, we plot the line $a_0 + a_1 x$ along with data, *and see what we expected: a line of best fit!*

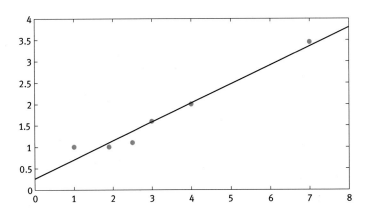

8.2.2 Curve of best fit

Since many data relations are not linear, it is important to also consider 'curves of best fit'. The ideas of the previous section can be extended to 'curves of best fit' without too much difficulty. There are three types of curves we will consider herein:
1. exponential: $y = be^{mx}$;
2. inverse: $y = \frac{1}{mx+b}$;
3. power: $y = bx^m$.

Each of these functions depend on two parameters, which we name m and b. It is our goal that, given data, we will find the m and b that provide a curve of best fit. From our construction, the ideas below can easily be extended to other types of two-parameter curves.

The procedure for finding curves of best fit is to
1. linearize the data through a transformation;
2. fit a line of best fit to the linearized data points;
3. un-transform the line to get a curve of best fit.

Step 1 of this procedure is the hardest, as it requires us to look at data and determine what kind of function that data might represent. If the data is exponential, then we would expect $(x, \log(y))$ to look linear. Similarly, if the data comes from an inverse function, then we would expect $(x, \frac{1}{y})$ to appear linear. Hence step 1 requires some educating guessing, then checking that the transformed data is indeed linear.

Step 2 of the procedure is clear. Once we have linearized data, we already know how to find the line of best fit for it.

Step 3 is to take the line of best fit for the transformed data, and turn it into a curve for the original data. How to do this depends on the transformation, but for the exponential case, we fit a line $\log(y) = a_0 + a_1 x$, then to un-transform, we raise e to both sides and get $y = e^{a_0 + a_1 x} = e^{a_0} e^{a_1 x}$. If we set $b = e^{a_0}$ and $m = a_1$, we now have our parameters for a curve of best fit.

The 'transformation to linear' for each of these data types is as follows:
1. If the data comes from $y = be^{mx}$, then logging both sides gives $\log(y) = \log(b) + mx$. Thus a plot of x vs. $\log(y)$ will look linear.
2. If the data comes from $y = \frac{1}{mx+b}$, then taking reciprocals of both sides gives $y^{-1} = mx + b$, so a plot of x vs. y^{-1} will look linear.
3. If the data comes from $y = bx^m$, then taking logs of both sides gives $\log(y) = \log(b) + m \log(x)$. Thus a plot of $\log(x)$ vs. $\log(y)$ will look linear.

A strategy for picking a function is to take an educated guess at what function that data might be coming from, then transform accordingly and see if the data is linear. If so, fit with that function, but if not, try a different one.

Example 50. *Fit a curve to the data points.*

0.2500	0.3662
0.5000	0.4506
0.7500	0.5054
1.0000	0.5694
1.2500	0.6055
1.5000	0.6435
1.7500	0.7426
2.0000	0.9068
2.2500	0.9393
2.5000	1.1297
2.7500	1.2404
3.0000	1.4441
3.2500	1.5313
3.5000	1.7706
3.7500	1.9669
4.0000	2.3129
4.2500	2.5123
4.5000	2.9238
4.7500	3.3070

Our first step is to plot the data. Doing so gives us

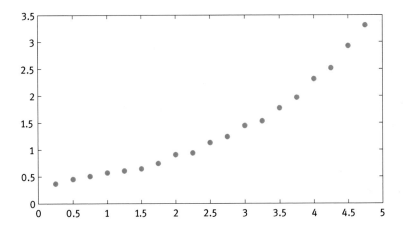

By examining the data, we suspect this could be either exponential or a power function.

Thus we plot the transformed data $(x, \log(y))$:

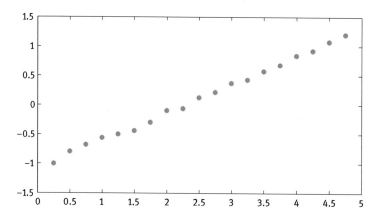

and $(\log(x), \log(y))$:

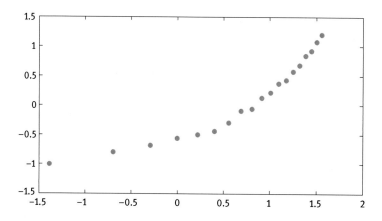

From these plots, we conclude the data most likely comes from an exponential function/distribution. The next step is to fit a line of best fit to the transformed linear data $(x, \log(y))$, *so use the line of best fit procedure, but using* $\log(y)$ *in place of y*

```
>> X = sum(x);
>> Y = sum(log(y));
>> W = sum( x.^2);
>> Z = sum( x.*log(y));
>> n = length(x);
>> A = [n X;X W];
>> b = [Y;Z];
>> a = A \ b
a =
    -1.0849
     0.4752
```

Thus we have fit a line of best fit to the data $(x_i, \log(y_i))$, and it given by

$$\log(y) = -1.0849 + 0.4752x$$

Now we un-transform that line by raising both sides to e to get

$$y = e^{-1.0849}e^{0.4752x} = 0.3379e^{0.4752x}$$

Since the line was a line of best fit for the transformed data, this will be curve of best fit for the original data, which we can see in the plot below.

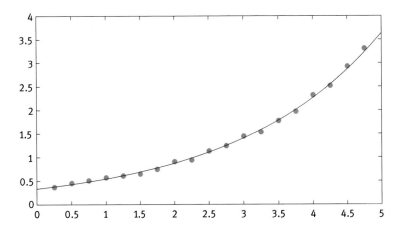

Let us now summarize the different curves to fit:

name:	equation:	fit line ($y = a_0 + a_1 x$) to:	answer:
line	$y = b + mx$	(x_i, y_i)	$b = a_0,\ m = a_1$
exponential	$y = b \cdot e^{mx}$	$(x_i, \log y_i)$	$b = e^{a_0},\ m = a_1$
power	$y = b \cdot x^m$	$(\log x_i, \log y_i)$	$b = e^{a_0},\ m = a_1$
inverse	$y = (b + mx)^{-1}$	$(x_i, 1/y_i)$	$b = a_0,\ m = a_1$

8.3 Exercises

1. Consider the data points

    x = [1   2   3   4   5   6   7   8   9   10]
    y = [1.1   4.1   8.9   16.1   26   36   52   63   78   105]

    (a) Find a single polynomial that interpolates the data points
    (b) Check with a plot that your polynomial does indeed interpolate the data
    (c) Would this interpolating polynomial likely give a good prediction of what is happening at $x = 1.5$? Why or why not?

(d) Create a cubic spline function for the same data, and plot it on the same graph as the single polynomial interpolant. Does it do a better job describing the data?

2. Consider the data points

```
x = [-1.0000         0    0.5000    1.0000    2.2500    3.0000    3.4000]
y = [0.3715    1.0849    1.7421    2.7862    9.5635   20.1599   30.0033]
```

(a) For the three possible choices from part the text, plot the corresponding transformed data, and decide which looks closest to linear.

(b) Find a curve of best fit for the data.

3. Consider the data points

0.1622	0.0043
0.7943	0.5011
0.3112	0.0301
0.5285	0.1476
0.1656	0.0045
0.6020	0.2181
0.2630	0.0182
0.6541	0.2798
0.6892	0.3274
0.7482	0.4188

Extend the ideas of this section for the 'line of best fit' to find the 'quadratic of best fit' $a_0 + a_1x + a_2x^2$ for this data. Hint: the procedure is similar, and the main difference is that now $e = e(a_0, a_1, a_2)$, and you will have 3 partials that must be zero.

4. Use the function $f(x) = \sin^2(x)$ on $[0, \pi]$ to verify the convergence theorems for piecewise linear and cubic spline interpolation. Using 11, 21, 41, and 81 equally spaced points, calculate the errors in the approximations (error as defined in the theorems), and then calculate convergence rates.

9 Numerical integration

In this chapter, we study methods of approximating the definite integral

$$\int_a^b f(x)dx.$$

The integrand, $f(x)$, is a function of x, and a and b are the bounds of integration. Many times, values for such integrals are found analytically, through antiderivatives and the fundamental theorem of Calculus. However, finding an antiderivative can be costly (i.e. takes time) or impossible, for example

$$\int_0^1 e^{x^2}\,dx.$$

In these cases, numerical integration is used to approximate the value of the integral.

9.1 Newton–Cotes methods

For Newton–Cotes integration methods, a value of the integrand at discrete points in the interval is estimated using a function that can be integrated easily. If the original integrand is in the form of data points, Newton–Cotes interpolates the integrand between the points.

Most students have already seen the common Newton–Cotes formulas in their calculus classes. These are shown in Table 9.1, and pictures of these rules for approximating the area under f on $[a, b]$ are shown below.

Table 9.1. Newton–Cotes quadrature rules.

Name	Rule for approximating $\int_a^b f(x)dx$
Left endpoint	$(b-a)f(a)$
Right endpoint	$(b-a)f(b)$
Midpoint	$(b-a)f\left(\frac{a+b}{2}\right)$
Trapezoid	$\frac{b-a}{2}\left(f(a) + f(b)\right)$
Simpson	$\frac{b-a}{6}f(a) + \frac{4(b-a)}{6}f(\frac{a+b}{2}) + \frac{b-a}{6}f(b)$

It is our goal to analyze these methods and determine which to use and when, and we proceed to do this using error analysis. It is possible to perform this analysis using Taylor series, however, it will be easier to instead look at what degree polynomials

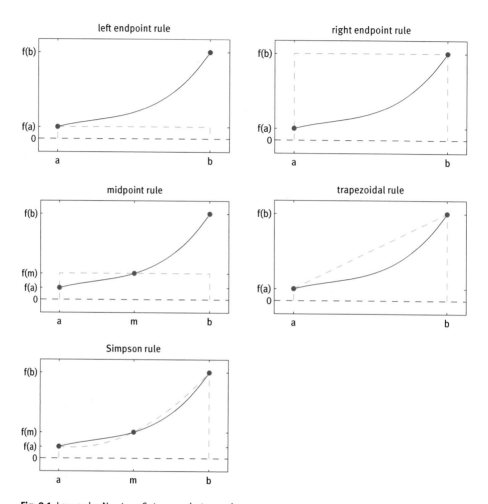

Fig. 9.1. Low order Newton–Cotes quadrature rules.

the rules are exactly correct for. The theorem below allows us to do this, as it shows this analysis technique is valid and will provide results equivalent to what Taylor series analysis would provide. However, we must be somewhat general here, so that the 'shortcut' applies to all of the rules. Hence, we define now the general quadrature rule on $[a, b]$ to be

$$Q(f) = \sum_{i=1}^{n} w_i f(x_i).$$

Here, the x_i's are the points where f is evaluated, and the w_i's are the weights. Note that each of the 5 Newton–Cotes quadrature rules from the table can be written in this form. For example:

- for the left endpoint, we have that $n = 1$, $x_1 = a$, and $w_1 = (b - a)$;
- for Simpson, $n = 3$, $x_1 = a$, $x_2 = \frac{a+b}{2}$, $x_3 = b$, $w_1 = w_3 = \frac{b-a}{6}$, and $w_2 = \frac{4(b-a)}{6}$.

We will also assume that the weights are positive, the rule is linear: $Q(f + g) = Q(f) + Q(g)$ and $Q(cf) = cQ(f)$, and $\sum_{i=1}^{n} w_i = b - a$. Note these assumptions are true for each of the Newton–Cotes rules.

Theorem 51. *Suppose a quadrature rule is exact on polynomials of degree m (i.e. $\int_a^b p_m(x)dx = Q(p_m)$ without any error at all). Then for $f \in C^{m+1}([a, b])$, the error in the quadrature rule is bounded by*

$$\left| \int_a^b f(x)dx - Q(f) \right| \le C(b - a)^{m+2}.$$

Proof. Write f in a Taylor expansion with expansion point a, as

$$f(x) = p_m(x) + \frac{f^{m+1}(c_x)}{(m + 1)!}(x - a)^{m+1},$$

with $x \le c_x \le a$, and where $p_m(x)$ is a degree m polynomial that is the leading terms of the Taylor expansion. Apply the quadrature rule to both sides, and using that Q is assumed to be linear, we get

$$Q(f) = Q\left(p_m + \frac{f^{m+1}(c_x)}{(m + 1)!}(x - a)^{m+1} \right) = Q(p_m) + \frac{f^{m+1}(c_x)}{(m + 1)!}Q\left((x - a)^{m+1} \right).$$

Since the rule is exact for polynomials of degree m, $Q(p_m) = \int_a^b p_m(x)dx$, and so

$$Q(f) = \int_a^b p_m(x)dx + \frac{f^{m+1}(c_x)}{(m + 1)!}Q\left((x - a)^{m+1} \right)$$

$$= \int_a^b \left(f(x) - \frac{f^{m+1}(c_x)}{(m + 1)!}(x - a)^{m+1} \right)dx + \frac{f^{m+1}(c_x)}{(m + 1)!}Q\left((x - a)^{m+1} \right)$$

$$= \int_a^b f(x)dx + \left(\frac{f^{m+1}(c_x)}{(m + 1)!}Q\left((x - a)^{m+1} \right) - \frac{f^{m+1}(c_x)}{(m + 1)!}\int_a^b (x - a)^{m+1}dx \right).$$

Thus we have isolated the error, and we have now, using the smoothness assumptions on f, that

$$\left| Q(f) - \int_a^b f(x)dx \right| = \left| \frac{f^{m+1}(c_x)}{(m + 1)!}Q\left((x - a)^{m+1} \right) - \frac{f^{m+1}(c_x)}{(m + 1)!}\int_a^b (x - a)^{m+1}dx \right|$$

$$\le \frac{\max_{a \le x \le b} |f^{m+1}(x)|}{(m + 1)!} \left(\left| Q\left((x - a)^{m+1} \right) \right| + \left| \int_a^b (x - a)^{m+1}dx \right| \right)$$

$$\leq C\left(\left| Q\left((x-a)^{m+1} \right) \right| + \left| \int_a^b (x-a)^{m+1} dx \right| \right)$$

$$\leq C\left(\left| Q\left((x-a)^{m+1} \right) \right| + (b-a)^{m+2} \right).$$

Now using the definition and assumptions on Q, it holds that

$$\left| Q\left((x-a)^{m+1} \right) \right| = \left| \sum_{i=1}^n w_i(x_i - a)^{m+1} \right| \leq \sum_{i=1}^n w_i |(x_i - a)^{m+1}| \leq \sum_{i=1}^n w_i(b-a)^{m+1} = (b-a)^{m+2}.$$

Combining the last two equations finishes the proof. □

This theorem makes it fairly easy to analyze the error from quadrature rules. We need to determine the highest degree m of polynomial for which a quadrature rule is exact, and then the error is $O((b-a)^{m+2})$ provided $f \in C^{m+1}([a, b])$.

Example 52. *Show the midpoint rule is exact on linears, and therefore for any* $f \in C^2([a, b])$, *its error satisfies*

$$\left| Q_{MP}(f) - \int_a^b f(x)dx \right| \leq C(b-a)^3.$$

First, we show the rule is exact on all constant functions. To do this, we let $f = 1$, *and calculate:*

$$Q_{MP}(f) = (b-a)f\left(\frac{a+b}{2} \right) = (b-a)(1) = (b-a)$$

$$\int_a^b f(x)dx = \int_a^b 1 dx = (b-a).$$

Note that both quadrature rules and integration are linear, and so this covers all constant functions.

Next, we show that the midpoint rule is exact on linears. Since we already showed this for constants, all that is left to do is show the rule is exact for $f(x) = x$:

$$Q_{MP}(f) = (b-a)f\left(\frac{a+b}{2} \right) = (b-a)\left(\frac{a+b}{2} \right) = \frac{b^2 - a^2}{2}$$

$$\int_a^b f(x)dx = \int_a^b x dx = \frac{b^2 - a^2}{2}.$$

Thus, the midpoint rule is exact on all linear functions. Now by the theorem, provided $f \in C^2([a, b])$, *it must hold that*

$$\left| Q_{MP}(f) - \int_a^b f(x)dx \right| \leq C(b-a)^3.$$

We can now analyze the error of each of the Newton–Cotes rules from above in this way. It is easy to see that the left and right endpoint rules will be exact only on constant functions, and the midpoint and trapezoid rules will be exact up to linears. Simpson's rule is exact on cubics, which we leave as an exercise. We put this information into Table 9.2.

Table 9.2. Newton–Cotes quadrature rules. The formula is exact for polynomials of degree m, which leads to the given error term as long as the given smoothness is satisfied.

Name	Rule	Exact for	Error on $[a, b]$	Required smoothness
Left endpoint	$(b - a)f(a)$	$m = 0$	$O((b - a)^2)$	$f \in C^1([a, b])$
Right endpoint	$(b - a)f(b)$	$m = 0$	$O((b - a)^2)$	$f \in C^1([a, b])$
Midpoint	$(b - a)f\left(\frac{a+b}{2}\right)$	$m = 1$	$O((b - a)^3)$	$f \in C^2([a, b])$
Trapezoid	$\frac{b-a}{2}(f(a) + f(b))$	$m = 1$	$O((b - a)^3)$	$f \in C^2([a, b])$
Simpson	$\frac{b-a}{6}f(a) + \frac{4(b-a)}{6}f(\frac{a+b}{2}) + \frac{b-a}{6}f(b)$	$m = 3$	$O((b - a)^5)$	$f \in C^4([a, b])$

9.2 Composite rules

Typically, it is not sufficient to use a single quadrature rule on $[a, b]$ to approximate $\int_a^b f(x)dx$. Instead, we divide $[a, b]$ into N intervals, and use a quadrature rule on each subinterval. Define the points $x_0, x_1, x_2, \ldots, x_N$ to be the x-points defining the subintervals, with $x_0 = a$ and $x_N = b$. Given a quadrature $Q(f, a, b)$ the corresponding composite quadrature Q^c is given by

$$Q^c(f, a, b) = \sum_{i=0}^{N-1} Q(f, x_i, x_{i+1}).$$

The composite rule for left rectangles is then

$$\int_a^b f(x)dx \approx f(x_0)(x_1 - x_0) + f(x_1)(x_2 - x_1) + f(x_2)(x_3 - x_2)$$
$$+ \cdots + f(x_i)(x_{i+1} - x_i) + \cdots + f(x_{N-1})(x_N - x_{N-1}).$$

For simplicity, in the remainder of this section we will assume equal spacing of the x-points, and we will call it $h = \frac{b-a}{N}$.

The composite Newton–Cotes rules can now be simplified as:
- composite left endpoint:
$$Q_L^c(f) = h \sum_{i=0}^{N-1} f(x_i);$$

- composite right endpoint:
$$Q_R^c(f) = h \sum_{i=1}^{N} f(x_i);$$

– composite midpoint:

$$Q_M^c(f) = h \sum_{i=0}^{N-1} f\left(\frac{x_i + x_{i+1}}{2}\right);$$

– composite trapezoidal:

$$Q_T^c(f) = \frac{h}{2} \sum_{i=0}^{N-1} (f(x_i) + f(x_{i+1}));$$

– composite Simpson:

$$Q_S^c(f) = \frac{h}{6} \sum_{i=0}^{N-1} \left(f(x_i) + 4f\left(\frac{x_i + x_{i+1}}{2}\right) + f(x_{i+1})\right).$$

We consider now the error in the composite rules. Here, the endpoints a and b are fixed, and let f be a given function. The theory we developed for these rules on a single interval can be used to make analyzing the composite rules' errors easy.

For each of the composite rules above, we can write

$$|error| = \left| \int_a^b f(x)dx - Q^c(f, a, b) \right| = \left| \sum_{i=0}^{N-1} \left(\int_{x_i}^{x_{i+1}} f(x)\, dx - Q(f, x_i, x_{i+1}) \right) \right|,$$

since the composite rule reduces to a single interval rule when we break up into individual intervals in the sum. Now consider the error on the single interval, interval i. The width of this interval is h, and thus the error is $O(h^k)$, where k depends on the rule being used (so $k = 2$ for left/right endpoint, 3 for midpoint/trap, and 5 for Simpson). This reduces the error bound to

$$|error| \le \sum_{i=0}^{N} Ch^k = C(N)h^k = C(b - a)h^{k-1},$$

since $(N - 1)h = (b - a)$. Since a and b are fixed, we have

$$|error| \le Ch^{k-1},$$

and we observe that the exponent from the single interval error is reduced by one for the composite rule error. We can now provide an error bound for each of the five rules, which is done in the table below.

Table 9.3. Newton–Cotes quadrature rules and associated errors.

Rule	exact on	Error on $[a, b]$, 1 subinterval	Error composite	needed smoothness
Left endpoint	constants	$O((b - a)^2)$	$O(h)$	$f \in C^1([a, b])$
Right endpoint	constants	$O((b - a)^2)$	$O(h)$	$f \in C^1([a, b])$
Midpoint	linears	$O((b - a)^3)$	$O(h^2)$	$f \in C^2([a, b])$
Trapezoid	linears	$O((b - a)^3)$	$O(h^2)$	$f \in C^2([a, b])$
Simpson	cubics	$O((b - a)^5)$	$O(h^4)$	$f \in C^4([a, b])$

Example 53. *Use the composite midpoint and Simpson rules with 10 subintervals to approximate $\int_1^2 \cos(x)dx$.*

We note first that the true solution is

```
>> truesoln = sin(2)-sin(1)

truesoln =

   0.067826442017785
```

To approximate it with the composite midpoint rule, we first write a function to do this in the general case. Note we assume equal point spacing.

```
function area = compMidpoint(f,a,b,n)
% function y = compMidpoint(f,a,b)
% Use composite midpoint to approx area under f on [a,b] with n
% subintervals

% create the n+1 points for n subintervals
x = linspace(a,b,n+1);
h = x(2)-x(1);

% Evaluate f at all the interval midpoints
fx = feval(f,(x(1:end-1) + x(2:end))/2);

% Apply midpoint formula
area = h * sum(fx);
```

Now we run the code and calculate the error.

```
>> area = compMidpoint(@cos,1,2,10)

area =

   0.067854711280263

>> midpointrule_error = truesoln - area

midpointrule_error =

   -2.826926247762040e-05
```

We repeat the process for Simpson. First, we write the code

```
function area = compSimpson(f,a,b,n)
% Use composite midpoint to approx area under f on [a,b] using n
% subintervals

% create the n+1 "node" points for n subintervals, and the midpoints
x = linspace(a,b,n+1);
xmid = (x(1:end-1) + x(2:end))/2;
```

```
h = x(2)-x(1);

% Rule is: sum( h/6 * (f(xi) + 4f(ximid) + f(xi+1)) )
fx = feval(f,x);
fxmids = feval(f,xmid);

area = h/6 * (fx(1) + fx(end) + 2*sum(fx(2:end-1)) + 4*sum(fxmids));
```

Then we run it.

```
>> area = compSimpson(@cos,1,2,10)

area =

   0.067826444373571

>> simprule_error = truesoln - area

simprule_error =

  -2.355785913565889e-09
```

It is clear in this example that the composite Simpson's method is much more accurate. Since the function is smooth, this is expected from the theory.

We now demonstrate the theoretical convergence rates proved for the composite midpoint rule do hold in practice.

Example 54. *Show that we get $O(h^2)$ convergence for the composite midpoint rule for the function $\sin(x) - x^3$ on $[1, 2]$.*

First, observe the true solution is

```
>> truesoln = cos(1)-cos(2)-(16-1)/4

truesoln =

  -2.793550857584718
```

Next, we write a MATLAB function for the given function.

```
function y = myfun4(x)
   y = sin(x) - x.^3;
```

To verify convergence rates, we run the composite midpoint rule with successively finer 'meshes' by cutting h in half repeatedly, then measuring the error.

```
>> area = compMidpoint(@myfun4,1,2,10);  error(1) = abs( truesoln - area);
>> area = compMidpoint(@myfun4,1,2,20);  error(2) = abs( truesoln - area);
>> area = compMidpoint(@myfun4,1,2,40);  error(3) = abs( truesoln - area);
>> area = compMidpoint(@myfun4,1,2,80);  error(4) = abs( truesoln - area);
>> area = compMidpoint(@myfun4,1,2,160); error(5) = abs( truesoln - area);
```

```
>> error'

ans =

   0.004148636741790
   0.001037137384174
   0.000259282983811
   0.000064820660826
   0.000016205159842

>> ratios = (error(1:end-1)./error(2:end) )'

ratios =

   4.000084082492218
   4.000021015379088
   4.000005253086067
   4.000001324005829
```

Thus we observe a factor of 4 between successive errors corresponding to cutting h in half. This is consistent with $O(h^2)$ convergence.

9.3 MATLAB's integral function

MATLAB has a built-in numerical integrator, called 'integral' (called 'quad' in older versions of MATLAB). It is used similar to the functions we wrote above. For example, to evaluate $\int_1^2 \cos(x)\, dx$, we use the syntax

```
>> integral(@cos,1,2)

ans =

   0.067826442017785
```

We can check that the answer is accurate to nearly machine precision.

```
>> sin(2)-sin(1)

ans =

   0.067826442017785
```

9.4 Gauss quadrature

One drawback of the composite methods we have seen is that the x-values are pre-elected. Sometimes, these points are not optimal. With Simpson's rule, for example, there are only three degrees of freedom in choosing coefficients. However, if we can choose the points ourselves, the degrees of freedom double, which allow us to achieve higher order accuracy with fewer points.

Consider the two point rule $I(f) \approx c_1 f(x_1) + c_2 f(x_2)$. This rule has four parameters, and we'd like to be exact for higher degree polynomials. Lets consider the interval $[-1, 1]$.

We first would like our parameters to be exact on constants, so let's consider $f(x) = 1$. Then

$$\int_{-1}^{1} 1 dx = c_1 + c_2 \rightarrow 2 = c_1 + c_2.$$

For linears, choose $f(x) = x$:

$$\int_{-1}^{1} x dx = c_1 x_1 + c_2 x_2 \rightarrow 0 = c_1 x_1 + c_2 x_2.$$

For quadratics, let $f(x) = x^2$:

$$\int_{-1}^{1} x^2 dx = c_1 x_1^2 + c_2 x_2^2 \rightarrow \frac{2}{3} = c_1 x_1^2 + c_2 x_2^2.$$

For cubics, let $f(x) = x^3$:

$$\int_{-1}^{1} x^3 dx = c_1 x_1^3 + c_2 x_2^3 \rightarrow 0 = c_1 x_1^3 + c_2 x_2^3.$$

This yields the following *nonlinear* system of four equations and four unknowns:

$$c_1 + c_2 = 2$$
$$c_1 x_1 + c_2 x_2 = 0$$
$$c_1 x_1^2 + c_2 x_2^2 = \frac{2}{3}$$
$$c_1 x_1^3 + c_2 x_2^3 = 0.$$

We now have to solve for the unknowns c_1, c_2, x_1, x_2. We know how to do this from our chapter on nonlinear solvers. Hence we write a functions for what we wish to find a zero of:

```
function y = Gauss2ptfun(x)
% call c1=x(1), c2 = x(2), x1 =x(3), x2=x(4)

y(1) = x(1) + x(2)-2;
y(2) = x(1)*x(3) + x(2)*x(4);
y(3) = x(1)*x(3)^2 + x(2)*x(4)^2 - 2/3;
y(4) = x(1)*x(3)^3 + x(2)*x(4)^3;
```

and its Jacobian

```
function y = Gauss2ptfunprime( x )

y(1,1)=1;
y(1,2)=1;
y(1,3:4)=0;

y(2,1) = x(3);
y(2,2) = x(4);
y(2,3) = x(1);
y(2,4) = x(2);

y(3,1) = x(3)^2;
y(3,2) = x(4)^2;
y(3,3) = 2 * x(1)*x(3);
y(3,4) = 2 * x(2)*x(4);

y(4,1) = x(3)^3;
y(4,2) = x(4)^3;
y(4,3) = 3 * x(1)*x(3)^2;
y(4,4) = 3 * x(2)*x(4)^2;
```

and then we call the n-dimensional Newton solver. A reasonable initial guess is that the points are the endpoints, and have equal weight. We run the code and get:

```
>> root = newtND(@Gauss2ptfun,@Gauss2ptfunprime,[1;1;-1;1],1e-12)

root =

    1.000000000000000
    1.000000000000000
   -0.577350269189626
    0.577350269189626
```

Note that $\frac{1}{\sqrt{3}} = 0.577350269189626$. Thus we obtain the solution $c_1 = c_2 = 1$, $x_1 = \frac{-1}{\sqrt{3}}$, $x_2 = \frac{1}{\sqrt{3}}$. So our quadrature rule is

$$Q_G^2(f, -1, 1) = f\left(\frac{-1}{\sqrt{3}}\right) + f\left(\frac{1}{\sqrt{3}}\right).$$

This is a two point rule which is exact on cubics, and so has accuracy of the same order as for Simpson's method. If we were to use a three point rule, derived in the same manner, it would be exact on quintics and require the solution of a nonlinear system with 6 equations and 6 unknowns (see exercises).

However, we assumed a region of integration of $[-1, 1]$, but we would like a rule for a general interval $[a, b]$. We can shift and scale the rule derived above to fit any region as follows:

$$Q_G^2(f, a, b) = \frac{b-a}{2}\left[f\left(\left(\frac{b-a}{2}\right)\left(\frac{-1}{\sqrt{3}}\right) + \left(\frac{a+b}{2}\right)\right)\right.$$
$$\left. + f\left(\left(\frac{b-a}{2}\right)\left(\frac{1}{\sqrt{3}}\right) + \left(\frac{a+b}{2}\right)\right)\right].$$

Example. Apply $\int_1^2 e^x dx$ with Gauss quadrature. Note that the true solution is 4.6708.

$$\int_1^2 e^x dx \approx \frac{1}{2}\left[f\left(\frac{-1}{2\sqrt{3}} + \frac{3}{2}\right) + \left(\frac{1}{2\sqrt{3}} + \frac{3}{2}\right)\right]$$

$$= \frac{1}{2}[f(1.2113) + f(1.7887)]$$

$$= \frac{1}{2}[3.3579 + 5.9815]$$

$$= 4.6697.$$

So the error in this rule is 0.0011. For Simpson, the error in this problem would be 0.0015. We expect that a two-point Gaussian rule is approximately equal to Simpson's rule in accuracy.

9.5 Exercises

1. Even though the midpoint and trapezoid rules have the same order of error, the midpoint rule usually gives about half the error of the trapezoid rule. Test this theory on the two functions.

$$f_1(x) = \sin^2(x), \quad f_2(x) = \cos(x)$$

    on the interval $[-1, 0.5]$.
2. Verify that the trapezoid rule is exact on all linears by showing it is exact for $f(x) = 1$ and $f(x) = x$.
3. Test the $O(h^4)$ convergence theory of the composite Simpson method for the function $f(x) = \sin^2(x)\cos(x)$ on $[0, 1]$. That is, successively cut h in half, and check that the error is being cut by a factor of 16.
4. Consider $f(x) = |\cos(x)|$ on $[0, 2]$. Test the accuracy of the composite left endpoint, midpoint, and Simpson methods on this function. Do your results agree with the theory? Why or why not?
5. Verify that Simpson methods is exact on cubics by checking that the rule is exact for $f(x) = 1, x, x^2, x^3$.
6. Write a program to compute 2 point Gauss quadrature on a single interval. Your input should be a function f, a, and b. Verify it by checking that the error in using the rule for $f(x) = x^3 - 3x^2 + x - 7$ is zero.

7. Write a program to compute a composite 2 point Gauss quadrature. The input should be a function f, a, b, and the number of subintervals to use n.
   Compare the accuracy of the 2-point Gauss quadrature on $f(x) = \sin^2(x)$ to composite Simpson's method, using the interval $[0, 2]$, with
   (a) 10 subintervals;
   (b) 100 subintervals.

8. Using the function $f(x) = \sin^2(x)$ on $[0, 2]$, verify that the 2 point Gauss quadrature rule converges with $O(h^4)$.

9. Derive the 3 points Gaussian rule that is exact on quintics, and verify that its accuracy is $O(h^6)$ on the test problem from the previous question.

10 Initial value ODEs

In this chapter, we take a deeper look at numerical methods for solving the initial value problem (a first order ODE)

$$y'(t) = f(t, y), \ t_0 \leq t \leq T, \qquad (10.1)$$
$$y(t_0) = y_0, \qquad (10.2)$$

where t_0 is the start time, T is the endtime, and y_0 is the initial condition. Note that even though we have written y and f as scalar functions, with all the methods we discuss below, they could be considered vectors functions (i.e. systems of first order ODEs) as well. A first order system of ODEs has the form:

$$y_1'(t) = f_1(t, y_1, \ldots, y_n)$$
$$y_2'(t) = f_2(t, y_1, \ldots, y_n)$$

$$\vdots \qquad \vdots$$

$$y_n'(t) = f_n(t, y_1, \ldots, y_n).$$

In many cases, closed form analytical solutions to initial value problems are not known, and even when they are, they may be very difficult to evaluate. In such cases, we resort to approximating solutions with numerical methods. The methods we discuss in this chapter will not look for an approximating function that is defined at every t in the interval $[t_0, T]$, but instead will approximate the function at some specified t-points, and from this we can interpolate the values at every t in the interval. That is, the 'solution' of a numerical ODE solver is a sequence of points

$$(t_0, y_0), \ (t_1, y_1), \ldots, (T, y_n).$$

For simplicity, we will consider equally spaced points, with spacing Δt.

We will show in the next section that solvers for this (seemingly simple) first order ODE above covers a very wide class of problems. This is because it covers vector systems, and higher order systems can usually be reduced to first order vector systems.

10.1 Reduction of higher order ODEs to first order

This section shows that many higher order ODEs can be written as vector systems of first order ODEs. Recall that the order of an ODE is the highest number of derivatives in any of its terms. For example, the ODE

$$y'''(t) + y(t)y''(t) - t^2 = 0$$

is a third order ODE. Provided the ODE can be written in the form

$$y^{(n)}(t) = F(t, y, y', \ldots, y^{(n-1)}),$$

then it can be written as a first order vector ODE by the following process:
- an nth order ODE will be turned into a first order ODE with n equations;
- define functions $u_1, u_2, \ldots, u_n$ by $u_1(t) = y(t)$ and $u_i(t) = y^{(i-1)}(t)$ for $i = 2, 3, \ldots, n$.
- The equations (identities) $u_i' = u_{i+1}$ for $i = 1, 2, \ldots, n-1$ form the first $n-1$ equations;
- for the last equation, use that

$$u_n' = y^{(n)}(t) = F(t, y, y', y'', \ldots, y^{(n-1)}) = F(t, u_1, u_2, u_3, \ldots, u_n).$$

Consider the following example.

Example 55. *Convert the following scalar ODE to a first order vector ODE:*

$$y'''(t) - ty''(t)y'(t) + y'(t) - ty(t) + \sin(t) = 0,$$
$$y(t_0) = y_0,$$
$$y'(t_0) = z_0,$$
$$y''(t_0) = w_0.$$

The first step is to create the u functions, and since this is a third order ODE, we need 3 of them:

$$u_1 = y,$$
$$u_2 = y',$$
$$u_3 = y''.$$

We now have the three equations in the u functions:

$$u_1' = u_2,$$
$$u_2' = u_3,$$
$$u_3' = tu_3 u_2 - u_2 + tu_1 - \sin(t).$$

Thus if we define

$$\mathbf{u}(t) = \begin{pmatrix} u_1(t) \\ u_2(t) \\ u_3(t) \end{pmatrix},$$

then we have the first order vector ODE

$$\mathbf{u}'(t) = F(t, \mathbf{u})$$

with initial condition

$$\mathbf{u}(t_0) = \begin{pmatrix} u_1(t_0) \\ u_2(t_0) \\ u_3(t_0) \end{pmatrix} = \begin{pmatrix} y(t_0) \\ y'(t_0) \\ y''(t_0) \end{pmatrix} = \begin{pmatrix} y_0 \\ z_0 \\ w_0 \end{pmatrix} = \mathbf{u}_0,$$

and

$$F(t, \mathbf{u}) = F(t, u_1, u_2, u_3) = \begin{pmatrix} u_2 \\ u_3 \\ tu_3 u_2 - u_2 + tu_1 - \sin(t) \end{pmatrix}.$$

10.2 Common methods and derivation from integration rules

Integrating the initial value ODE (10.1) from t_i to t_{i+1} gives

$$y(t_{i+1}) - y(t_i) = \int_{t_i}^{t_{i+1}} f(s, y(s)) \, ds. \tag{10.3}$$

Suppose we applied the left endpoint rule to approximate the integral. This would yield

$$y(t_{i+1}) - y(t_i) \approx (t_{i+1} - t_i)f(t_i, y(t_i)).$$

Hence if we knew $y(t_i)$ (or some approximation of it), we could use this formula to approximate $y(t_{i+1})$. We now proceed by changing the approximation sign to an equals sign, and replace the true solution $y(t_i)$ with its approximation y_i. This procedure can be formalized as the forward Euler formula, which we derived earlier in this book from finite difference methods.

Algorithm 56 (Forward Euler). *Given a discretization in time t_0, t_1, t_2, ..., $t_n = T$ and initial condition y_0, find y_1, y_2, ..., y_n by the following sequential process:*

$$y_{i+1} = y_i + \Delta t f(t_i, y_i).$$

A nice property of forward Euler is that it is simple and *explicit* (explicit in the sense that you plug in known values t_i and y_i into the function f and no equation needs to be solved). However, it is also not very accurate and can have stability issues, as we will discuss in this chapter. We will not study accuracy of ODE solvers in detail; however, a good rule of thumb is that the ODE solver is as accurate as the approximations used to create it. Since forward Euler is created using the left endpoint rule, and we integrate from t_0 to T on many small intervals, then the error in forward Euler comes from the error in composite left endpoint quadrature, which is first order, and thus error is $O(\Delta t)$. To be more specific, for forward Euler it holds that

$$\max_{1 \le i \le n} \left| y(t_i) - y_i^{FE} \right| \le C\Delta t,$$

where $y(t_i)$ is the true solution at t_i. The MATLAB code for forward Euler is given below.

```
function y = forwardEuler(func,t,y1)
% y = forwardEuler(func,t,y1)
% solve the ODE y'=f(t,y) with initial condition y(t1)=y1 and return
% the function values as a vector.
% func is the function handle for f(t,y) and t is a vector
% of the times: [t1,t2,...,tn].

% N is the total number of points, initialize y to be the same
% size as t:
```

```
N = length(t);
y = zeros(N,1);

% Set initial condition:
y(1)=y1;

% use forward Euler to find y(i+1) using y(i)
for i=1:N-1
    y(i+1) = y(i) + ( t(i+1) - t(i) ) * func(t(i),y(i));
end
```

10.2.1 Backward Euler

Suppose instead of the left endpoint rule, we applied the right endpoint rule to approximate the integral in (10.3). Then we would get

$$y(t_{i+1}) - y(t_i) \approx \Delta t f(t_{i+1}, y(t_{i+1})).$$

So again, if we knew $y(t_i)$ or some approximation of it, then we could use this formula to approximate what $y(t_{i+1})$ is. This defines the backward Euler formula.

Algorithm 57 (backward Euler). *Given a discretization in time $t_0, t_1, t_2, \ldots, t_n = T$ and initial condition y_0, find $y_1, y_2, \ldots, y_n$ by the following sequential process:*

$$y_{i+1} = y_i + \Delta t f(t_{i+1}, y_{i+1}).$$

The obvious disadvantage to backward Euler is that it is an *implicit* method, as it must perform a solve to find y_{i+1}, whereas in forward Euler there is no solve – just an explicit calculation. However, as we will see in the next section, an advantage of backward Euler is that it is unconditionally stable. Since it is derived from the right endpoint rule, its accuracy is also $O(\Delta t)$. A code for backward Euler is given below.

```
function y = backwardEuler(func,t,y1)
% y = backwardEuler(func,t,y1)
% solve the ODE y'=f(t,y) with initial condition y(t1)=y1 and return
% the function values as a vector.
% func is the function handle for f(t,y) and t is a vector
% of the times: [t1,t2,...,tn].

% N is the total number of points, initialize y to be the same
% size as t:
N = length(t);
y = zeros(N,1);

% initial condition
y(1)=y1;
```

```
% use BE to find y_i+1
for i=1:length(y)-1
    % solve y(i+1) - y(i) + dt * f(t(i+1),y(i+1)) = 0
    % define an inline function where y(i+1) is independent variable
    ode_eqn = @(ynext) ynext - y(i) - ( t(i+1) - t(i) )...
                 * feval(func,t(i+1),ynext);
    % solve w/ Matlab's 1d solver, guess y(i) is the solution
    y(i+1) = fzero(ode_eqn,y(i));
end
```

10.2.2 Crank–Nicolson

If we apply the midpoint rule to (10.3) and the approximation

$$y\left(\frac{t_i + t_{i+1}}{2}\right) \approx \frac{1}{2}\left(y(t_i) + y(t_{i+1})\right),$$

we get

$$y(t_{i+1}) - y(t_i) \approx \Delta t f\left(\frac{t_{i+1} + t_i}{2}, \frac{y_i + y_{i+1}}{2}\right).$$

This approximation defines the Crank–Nicolson solver for initial value ODEs.

Algorithm 58 (Crank–Nicolson). *Given a discretization in time t_0, t_1, $t_2, \ldots, t_n = T$ and initial condition y_0, find y_1, $y_2, \ldots, y_n$ by the following sequential process:*

$$y_{i+1} = y_i + \Delta t f\left(\frac{t_{i+1} + t_i}{2}, \frac{y_i + y_{i+1}}{2}\right).$$

Similar to backward Euler, this method is implicit, and we will show that this method is also unconditionally stable. This method is more accurate than forward and backward Euler, which is a result of its derivation coming from a more accurate quadrature approximation - its error is $O(\Delta t^2)$. We leave the creation of a Crank–Nicolson code, which is very similar to backward Euler code, as an exercise.

10.2.3 Runge–Kutta 4

The last method we will formally discuss is fourth order Runge Kutta (often called RK4). It is derived by applying the Simpson quadrature formula to (10.3), then rewriting it in a clever way. It is defined in four, explicit, steps:

Algorithm 59 (RK4). *Given a discretization in time t_0, t_1, $t_2, \ldots, t_n = T$ and initial condition y_0, let $\Delta t_i = (t_{i+1} - t_i)$, and find y_1, $y_2, \ldots, y_n$ by the following sequential process:*

$$K_1 = f(t_i, y_i),$$

$$K_2 = f\left(t_i + \frac{\Delta t_i}{2}, y_i + \frac{\Delta t_i}{2}K_1\right),$$

$$K_3 = f\left(t_i + \frac{\Delta t_i}{2}, y_i + \frac{\Delta t_i}{2}K_2\right),$$

$$K_4 = f\left(t_i + \Delta t_i, y_i + \Delta t_i K_3\right),$$

$$y_{i+1} = y_i + \Delta t_i \left(\frac{K_1}{6} + \frac{K_2}{3} + \frac{K_3}{3} + \frac{K_4}{6}\right).$$

RK4 is explicit and therefore very fast, but it is significantly more accurate than the other methods above. It can have stability issues, however, but when it works (which is usually), it is hard to beat. It is probably the most commonly used ODE solver for initial value problems. Its accuracy is $O(\Delta t^4)$, since it is derived from Simpson's rule. The MATLAB code for RK4 is given below.

```
function y = rk4(func,t,y1)
% function y = rk4(func,t,y1)
% compute approx to dy/dt = f(t,y) with initial condition y(t1)=y1
% t is a vector of the t points: [t1,t2,...,tn]

% initialize y to be the same size at t
y = 0 * t;

% initial condition
y(1)=y1;

for i=1:length(y)-1
    dt = t(i+1) - t(i);

    K1 = feval(func, t(i), y(i));
    K2 = feval(func, t(i) + dt/2, y(i) + dt/2*K1);
    K3 = feval(func, t(i) + dt/2, y(i) + dt/2*K2);
    K4 = feval(func, t(i) + dt  , y(i) + dt  *K3);

    y(i+1) = y(i) + dt * ( K1/6 + K2/3 + K3/3 + K4/6 );

end
```

10.3 Comparison of speed of implicit versus explicit solvers

A calculation of a timestep of forward Euler requires us to **calculate**

$$y_{i+1} = y_i + \Delta t f(t_i, y_i).$$

Likely, the evaluation of f will be the dominant cost here.

A calculation of a timestep using backward Euler requires us to **solve**

$$y_{i+1} = y_i + \Delta t f(t_{i+1}, y_{i+1}).$$

Likely Newton's method is the best choice to solve this system (of course, for simpler problems, something else might be possible). To use Newton's method, we need to set up the problem as

$$g(x) = 0,$$

which can be done by defining

$$g(x) = x - y_i - \Delta t f(t_{i+1}, x).$$

Then the Newton iteration becomes

$$x_{k+1} = x_k - \frac{g(x_k)}{g'(x_k)} = \frac{x_k - y_i - \Delta t f(t_{i+1}, x_k)}{1 - \Delta t f_x(t_{i+1}, x_k)}$$

in the scalar case, and

$$x_{k+1} = x_k - J^{-1} g(x_k)$$

in the vector case, where J is the Jacobian of g with respect to x.

Each Newton iteration requires 1 evaluation of f, 1 evaluation of the derivative of f, and 1 'solve' (in the scalar case, the solve is just a division). Since Newton often requires at least 4 iterations to converge, this means the total cost of backward Euler is (4 evaluation of f) + (4 evaluations of derivative of f) + 4 solves.

In the scalar case, the solve is negligible, and since the cost of evaluation of derivatives is usually at least as expensive as function evaluations, the cost of backward Euler is at least 8 times slower than forward Euler! In the vector case, evaluating the derivative becomes more expensive, since the Jacobian is now an $n \times n$ matrix. Hence evaluating the derivative here is typically n times as costly as evaluating the function. Moreover, doing the linear solve is a significant added cost, and if $n > 100$, this will be the dominant cost. Even ignoring the solve, we get that backward Euler will take $(4 + 4n)$ times longer than forward Euler.

This type of slowdown is typical of implicit methods (ones that require a solve) compared to explicit methods (ones that are just a calculation). The reason implicit methods are used at all is that they are more stable than explicit methods, which is discussed in detail in the next section. In short, this means they are less likely to "blow up" and give garbage answers for seemingly no reason.

For a test example, we choose the initial value problem

$$y'(t) = -1.2y + 7e^{-0.3t}, \quad y(0) = 3, \quad 0 \le t \le 2.5 = T,$$

and thus we write MATLAB code below for the right hand side function.

We run each method using $\Delta t = T/1000$, and time it using tic and toc, which gives us the following output

```
>> t = linspace(0,2.5,1001);
>> tic; y = rk4(@odefun1,t,3); toc
Elapsed time is 0.037007 seconds.

>> tic; y = forwardEuler(@odefun1,t,3); toc
Elapsed time is 0.009678 seconds.
```

```
>> tic; y = backwardEuler(@odefun1,t,3); toc
Elapsed time is 0.775298 seconds.

>> tic; y = CrankNicolson(@odefun1,t,3); toc
Elapsed time is 0.777697 seconds.
```

Thus we see that even for this simple scalar problem, the implicit methods are dramatically slower. Notice that RK4 is about 4 times slower than forward Euler, which is not surprising since RK4 using 4 function evaluations and forward Euler uses 1.

10.4 Stability of ODE solvers

The notion of 'stability' of an ODE solver is based on the notion that 'if the ODE solution does not exhibit exponential growth, then neither should the numerical method'. Recall from calculus that if we 'zoom in' close enough on a smooth function, it looks linear. Moreover, exponential growth of an ODE is caused by a dependence of f on y, not t. Hence, locally, it is reasonable to represent an ODE of the form

$$y' = f(t, y),$$

by

$$y' = \lambda y.$$

Since we know the solution to this ODE is $y(t) = e^{\lambda t}$, it is easy to see that exponential growth occurs only if $\text{Re}(\lambda) > 0$. We will assume in this section that the ODE itself is stable, i.e. we will assume that $\text{Re}(\lambda) \leq 0$. Our concern with numerical methods will then be whether or not they exhibit exponential growth when the ODE does not.

10.4.1 Stability of forward Euler

For the ODE $y' = \lambda y$, the forward Euler method takes the form

$$y_{n+1} = y_n + \Delta t f(t_n, y_n) = y_n + \Delta t \lambda y_n = (1 + \lambda \Delta t) y_n = \cdots = (1 + \lambda \Delta t)^{n+1} y_0.$$

Hence forward Euler does not exhibit exponential growth if $|1 + \lambda \Delta t| \leq 1$, i.e. if $|\lambda| \Delta t < 2$. Since λ is a complex number, this region can be considered a circle of radius 1, centered at $(-1, 0)$ in the complex plane. A simple way to think of this is: the larger λ is, the smaller Δt needs to be.

Example 60. *A chemical decays proportional to its concentration to the 1.5 power, and simultaneously it is being produced. The ODE for its concentration is given by*

$$y' = -0.8 * y^{1.5} + 20000(1 - e^{-3t}).$$

If $y(0) = 2000$, approximate y on $[0, 0.5]$ with $\Delta t = 0.1, 0.05, 0.01, 0.001$.
We write a right hand side function

```
function f = odefun2(t,y)

f = -0.8*y.^(1.5) + 20000*(1-exp(-3*t));
```

Then we run it and plot the solution for the various Δt's using the MATLAB commands

```
>> t = linspace(0,.5,6);
>> y = forwardEuler(@odefun2,t,2000);
>> plot(t,y,'k.')

>> t = linspace(0,.5,11);
>> y = forwardEuler(@odefun2,t,2000);
>> plot(t,y,'k.')

>> t = linspace(0,.5,51);
>> y = forwardEuler(@odefun2,t,2000);
>> plot(t,y,'k.')

>> t = linspace(0,.5,501);
>> y = forwardEuler(@odefun2,t,2000);
>> plot(t,y,'k.')
```

The plots we get are shown in Figure 10.1. It is clear to see (look at the scale of the y-axis) that we experience unstable behavior (blow up to infinity) for $\Delta t = 0.1$ and $\Delta t = 0.05$ solutions. **Even though forward Euler converges, until Δt is small enough for it to be stable, the answer is no good.** *Once Δt is small enough, then we see $O(\Delta t)$ convergence to the solution.*

10.4.2 Stability of backward Euler

Consider now the stability of backward Euler applied to the ODE $y' = \lambda y$. Using the backward Euler algorithm, we get

$$y_{n+1} = y_n + \Delta t \lambda y_{n+1},$$

and so

$$y_{n+1} = (1 - \lambda \Delta t)^{-1} y_n = \cdots = (1 - \lambda \Delta t)^{-(n+1)} y_0.$$

Thus exponential growth happens only if $\left| \frac{1}{1-\lambda \Delta t} \right| > 1$. However, since $Re(\lambda) > 0$, the denominator has a real part bigger than 1, for any $\Delta t > 0$. Thus the modulus must al-

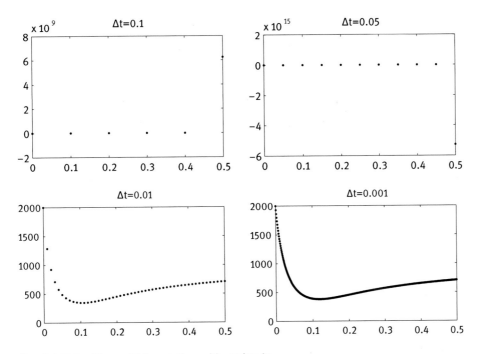

Fig. 10.1. Plots of forward Euler solutions with varying Δt.

ways be bigger than 1. Hence there is no restriction on Δt for backward Euler to be stable, and thus we say **backward Euler is unconditionally stable.** Consider the same example as above for forward Euler, now use backward Euler.

Example 61. *Repeat Example 60, but now use backward Euler.*
We run the backward Euler code and plot the solution for the various Δt's using the MATLAB commands

```
>> t = linspace(0,.5,6);
>> y = backwardEuler(@odefun2,t,2000);
>> plot(t,y,'k.')

>> t = linspace(0,.5,11);
>> y = backwardEuler(@odefun2,t,2000);
>> plot(t,y,'k.')

>> t = linspace(0,.5,51);
>> y = backwardEuler(@odefun2,t,2000);
>> plot(t,y,'k.')

>> t = linspace(0,.5,501);
>> y = backwardEuler(@odefun2,t,2000);
>> plot(t,y,'k.')
```

The plots we get are shown below in Figure 10.2. From the plots, we see that backward Euler did not blow up for the smaller Δt's, as forward Euler did. For smaller timesteps, the solution matches that of the (stable) forward Euler method well.

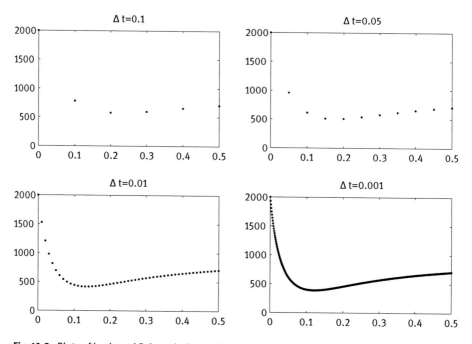

Fig. 10.2. Plots of backward Euler solutions with varying Δt.

10.4.3 Stability of Crank–Nicolson

Applying Crank–Nicolson to $y' = \lambda y$ gives

$$y_{n+1} = y_n + \Delta t f\left(\frac{t_n + t_{n+1}}{2}, \frac{y_n + y_{n+1}}{2}\right) = y_n + \lambda\frac{\Delta t}{2}(y_n + y_{n+1}).$$

Grouping terms and isolating the y_{n+1} gives

$$y_{n+1} = \left(\frac{1 + \frac{\lambda \Delta t}{2}}{1 - \frac{\lambda \Delta t}{2}}\right)y_n = \cdots = \left(\frac{1 + \frac{\lambda \Delta t}{2}}{1 - \frac{\lambda \Delta t}{2}}\right)^{n+1}y_0.$$

Since we assume that $\text{Re}(\lambda) < 0$, then the real part of the denominator must be bigger than the real part of the numerator. Further, the complex parts of the numerator and denominator have the same magnitude. Therefore, we have that

$$\left|\left(\frac{1 + \frac{\lambda \Delta t}{2}}{1 - \frac{\lambda \Delta t}{2}}\right)\right| < 1$$

for any $\Delta t > 0$. **Thus we say Crank–Nicolson is unconditionally stable.**

10.4.4 Stability of Runge–Kutta 4

As you might expect from the formula, analyzing the stability of RK4 is significantly more involved than the methods we analyzed stability for above. Indeed, the same procedure is used, but the algebra gets messy. Hence we just state the result: Stability of RK4 requires that

$$\left| 1 + \lambda \Delta t + \frac{1}{2}(\lambda \Delta t)^2 + \frac{1}{6}(\lambda \Delta t)^3 + \frac{1}{24}(\lambda \Delta t)^4 \right| \leq 1.$$

This ends up being a similar conditional stability result as for forward Euler.

10.5 Accuracy of ODE solvers

We have stated above that by their construction, we expect that the forward and backward Euler solvers have accuracy $O(\Delta t)$, Crank–Nicolson is $O(\Delta t^2)$, and RK4 is $O(\Delta t^4)$. Derivation of these errors using analysis is beyond the scope of this course, however we can at least try to verify that these results are true for a test problem.

We choose the initial value problem

$$y'(t) = -1.2y + 7e^{-0.3t}, \quad y(0) = 3, \quad 0 \leq t \leq 2.5 = T.$$

So that we can compare accuracies, we will compute errors using the true solution

$$y(t) = \frac{70}{9}e^{-0.3t} - \frac{43}{9}e^{-1.2t}.$$

We will compute each of the methods with Δt = T/10, T/20, T/40 and T/80, and calculate the errors, with error defined by

$$err = \max_{i=1,2,\dots,n} \left| y_{true}(t_i) - y_i \right|$$

We first write the MATLAB right hand side function 'odefun1', which is shown below

```
function f = odefun1(t,y)
    f = -1.2*y + 7*exp(-0.3*t);
```

10.5.1 Forward Euler

We begin with forward Euler. The MATLAB commands are shown below, and we observe that the error decreases by a factor of about half each time that Δt is cut in half. This is consistent with $O(\Delta t)$ accuracy.

```
>> t = linspace(0,2.5,11);
>> y = forwardEuler(@odefun1,t,3);
>> ytrue = 70/9 * exp(-0.3*t) - 43/9 * exp(-1.2*t);
>> max(abs(y - ytrue))

ans =

    0.2610

>> t = linspace(0,2.5,21);
>> ytrue = 70/9 * exp(-0.3*t) - 43/9 * exp(-1.2*t);
>> y = forwardEuler(@odefun1,t,3);
>> max(abs(y - ytrue))

ans =

    0.1205

>> t = linspace(0,2.5,41);
>> ytrue = 70/9 * exp(-0.3*t) - 43/9 * exp(-1.2*t);
>> y = forwardEuler(@odefun1,t,3);
>> max(abs(y - ytrue))

ans =

    0.0580

>> t = linspace(0,2.5,81);
>> ytrue = 70/9 * exp(-0.3*t) - 43/9 * exp(-1.2*t);
>> y = forwardEuler(@odefun1,t,3);
>> max(abs(y - ytrue))

ans =

    0.0285
```

10.5.2 Backward Euler

Next, we consider backward Euler. The MATLAB commands are shown below, and we observe that the error decreases by a factor of about half each time that Δt is cut in half. This is consistent with $O(\Delta t)$ accuracy.

```
>> t = linspace(0,2.5,11);
>> y = backwardEuler(@odefun1,t,3);
>> ytrue = 70/9 * exp(-0.3*t) - 43/9 * exp(-1.2*t);
>> max(abs(y - ytrue))
```

```
ans =

    0.1975

>> t = linspace(0,2.5,21);
>> y = backwardEuler(@odefun1,t,3);
>> ytrue = 70/9 * exp(-0.3*t) - 43/9 * exp(-1.2*t);
>> max(abs(y - ytrue))

ans =

    0.1049

>> t = linspace(0,2.5,41);
>> y = backwardEuler(@odefun1,t,3);
>> ytrue = 70/9 * exp(-0.3*t) - 43/9 * exp(-1.2*t);
>> max(abs(y - ytrue))

ans =

    0.0542

>> t = linspace(0,2.5,81);
>> ytrue = 70/9 * exp(-0.3*t) - 43/9 * exp(-1.2*t);
>> y = backwardEuler(@odefun1,t,3);
>> max(abs(y - ytrue))

ans =

    0.0276
```

10.5.3 Crank–Nicolson

Next we test Crank–Nicolson. The MATLAB commands are shown below, and we observe that the error decreases by a factor of 4 each time that Δt is cut in half. This is consistent with $O(\Delta t^2)$ accuracy. Also note the actual errors are much smaller than for forward and backward Euler.

```
>> t = linspace(0,2.5,11);
>> y = CrankNicolson(@odefun1,t,3);
>> ytrue = 70/9 * exp(-0.3*t) - 43/9 * exp(-1.2*t);
>> max(abs(y - ytrue))

ans =

    0.0106

>> t = linspace(0,2.5,21);
>> y = CrankNicolson(@odefun1,t,3);
```

```
>> ytrue = 70/9 * exp(-0.3*t) - 43/9 * exp(-1.2*t);
>> max(abs(y - ytrue))

ans =

    0.0026

>> t = linspace(0,2.5,41);
>> y = CrankNicolson(@odefun1,t,3);
>> ytrue = 70/9 * exp(-0.3*t) - 43/9 * exp(-1.2*t);
>> max(abs(y - ytrue))

ans =

    6.5621e-04

>> t = linspace(0,2.5,81);
>> y = CrankNicolson(@odefun1,t,3);
>> ytrue = 70/9 * exp(-0.3*t) - 43/9 * exp(-1.2*t);
>> max(abs(y - ytrue))

ans =

    1.6397e-04
```

10.5.4 Runge–Kutta 4

Next we test RK4. The MATLAB commands are shown below, and we observe that the error decreases by a factor of about 16 each time that Δt is cut in half. This is consistent with $O(\Delta t^4)$ accuracy. Also note the actual errors are much smaller than the other three methods.

The MATLAB code for RK4 is shown below.

```
>> t = linspace(0,2.5,11);
>> y = rk4(@odefun1,t,3);
>> ytrue = 70/9 * exp(-0.3*t) - 43/9 * exp(-1.2*t);
>> max(abs(y - ytrue))

ans =

    1.2804e-04

>> t = linspace(0,2.5,21);
>> y = rk4(@odefun1,t,3);
>> ytrue = 70/9 * exp(-0.3*t) - 43/9 * exp(-1.2*t);
>> max(abs(y - ytrue))
```

```
ans =

   7.0050e-06

>> t = linspace(0,2.5,41);
>> y = rk4(@odefun1,t,3);
>> ytrue = 70/9 * exp(-0.3*t) - 43/9 * exp(-1.2*t);
>> max(abs(y - ytrue))

ans =

   4.0967e-07

>> t = linspace(0,2.5,81);
>> y = rk4(@odefun1,t,3);
>> ytrue = 70/9 * exp(-0.3*t) - 43/9 * exp(-1.2*t);
>> max(abs(y - ytrue))

ans =

   2.4773e-08
```

10.6 Summary, general strategy, and MATLAB ODE solvers

We have the following results for the four methods we discussed.

Method	speed	stability	accuracy
forward Euler	fast	conditional	$O(\Delta t)$
backward Euler	slow	unconditional	$O(\Delta t)$
Crank–Nicolson	slow	unconditional	$O(\Delta t^2)$
Runge–Kutta 4	fast	conditional	$O(\Delta t^4)$

Note from the table and discussion above, one should almost always choose RK4 instead of forward Euler, and similarly one should almost always choose Crank–Nicolson over backward Euler.

Hence a general strategy for solving ODE's is the following: Try an explicit method such as RK4 first. If it works, you are done! If it has stability issues, you will have to resort to an implicit method. Of the implicit methods, Crank–Nicolson is often a good choice, since it is $O(\Delta t^2)$ accurate, whereas backward Euler is only $O(\Delta t)$.

Of course, there are many more methods for solving ODEs than the four above. But now that we understand the basics, in most cases we can simply use the existing ODE solvers in MATLAB. The MATLAB solver 'ode45' is similar to RK4, but has numerous improvements built into it. For example, it shrinks the timestep as necessary to get stability and a desired accuracy. Hence ode45 is still an explicit (fast) and highly accurate solver provided the timestep restriction for stability is not too restrictive.

For problems where stability can still be an issue with ode45 (such ODEs are called 'stiff'), the solver ode15s should be tried next. It is an implicit solver, that switches between different methods so that it can achieve optimal accuracy while remaining stable. Note that for any implicit method, a solve has to be done, and so MATLAB lets you provide the derivative/Jacobian of the right hand side function f. If you don't provide it, MATLAB will use finite differences, which in some cases can be inaccurate enough to cause the method to fail.

Let's do an example.

Example 62. *Solve the initial value problem on $1 \leq t \leq 2$:*

$$t^3 y''' - t^2 y'' + 3ty' - 4y = 5t^3 \ln t + 9t^2$$
$$y(1) = 0$$
$$y'(1) = 1$$
$$y''(1) = 3.$$

First, we convert the ODE to first order. This gives the system

$$\begin{pmatrix} u_1 \\ u_2 \\ u_3 \end{pmatrix}'(t) = \mathbf{u}'(t) = \mathbf{g}(t, \mathbf{u}) = \begin{pmatrix} u_2 \\ u_3 \\ t^{-1} u_3 - 3t^{-2} u_2 + 4t^{-3} u_1 + 5 \ln t + 9t^{-1} \end{pmatrix}$$

with initial condition

$$\mathbf{u}(1) = \begin{pmatrix} 0 \\ 1 \\ 3 \end{pmatrix}.$$

Note that dividing through by the t^3 was permissible since it is never 0 on the domain.
Next, we create a right hand side function f to pass into MATLAB.

```
function f = odefun3(t,u)

f(1,1) = u(2);
f(2,1) = u(3);
f(3,1) = u(3)/t - 3*u(2)/(t^2) + 4*u(1)/(t^3) + 5*log(t) + 9/t;
```

The ode solver can then be called with

```
>> [t,y] = ode45(@odefun3,[1 2],[0 1 3]);
```

Notice that we only gave MATLAB the start and end time, we did not discretize t. MATLAB uses its defaults to pick each individual timestep, and they are not (usually) chosen uniformly. In this case, there were 57 t points chosen

```
>> size(t)

ans =

    57      1
```

However, we are allowed to increase the desired accuracy with the odeset options. The default absolute error tolerance is 1e-6, and if we make it smaller, then more t points are used.

```
>> options = odeset('AbsTol',1e-10);
>> [t,y] = ode45(@odefun3,[1 2],[0 1 3],options);
>> size(t)

ans =

    81      1
```

The solution is held in y, but note that y holds all 3 components: $[y, y', y'']$. If we want to plot just y, we say

```
>> plot(t,y(:,1),'k.-')
```

to produce the plot

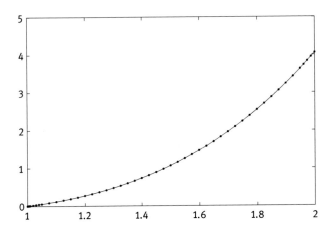

10.7 Exercises

1. Consider the following initial value problem on $0 \le t \le 2$:

    $$y'(t) = e^t (\sin(t) + \cos(t)) + 2t, \quad y(0) = 1.$$

    Calculate solutions to this problem using forward Euler, backward Euler, Crank–Nicolson and RK4, each with varying discretizations $\Delta t = \frac{1}{10}, \frac{1}{20}, \frac{1}{40}, \frac{1}{80}, \frac{1}{160}$.

Create a table of errors and convergence rates, noting that the true solution is

$$y(t) = e^t \sin(t) + t^2 + 1.$$

Do your estimated convergence rates match the theory for each method? Note that you need to create your own Crank–Nicolson code, and it would be a good idea to create it by changing the backward Euler code.

2. Solve the following initial value problem using ode45 and ode15s:

$$y'''(t) - 3y''(t) + ty(t) - \sin^2(t) = 7, \quad 0 \le t \le 1, \quad y(0) = 0, \quad y'(0) = 1, \quad y''(0) = 0.$$

Plot the solution for varying tolerances. Why do you believe your solution is correct?

3. Consider the initial value problem on $0 \le t \le 3$:

$$\begin{pmatrix} u_1 \\ u_2 \end{pmatrix}' = \begin{pmatrix} -4u_1 + 2u_2 + \cos(t) + 4\sin(t) \\ 3u_1 + u_2 - 3\sin(t) \end{pmatrix}.$$

The initial condition is $u_1(0) = 1$, $u_2(0) = 1$. Solve this ODE using MATLAB's ode15s and ode45s. Plot your solution. Why do you believe your solution is correct?

4. Water flows out of an inverted conical talk with circular orifice at the rate of

$$y'(t) = -.7\pi r^2 \sqrt{-2g} \frac{\sqrt{y}}{A(y)},$$

where r is the orifice radius, y is the height of the water above the orifice, and $A(y)$ is the area of the cross section of the tank at the water level. Suppose $r = .1$ ft, $g = -32.17$ ft/s, and the tank has an initial water level of 8 ft and initial volume of $512(\pi/3)$ ft^3. Use ode45 to approximate a solution, and use it to determine when the tank will be empty.

A. Getting started with Octave and MATLAB

This book uses the Octave/MATLAB computing language and commands. While we expect students to have some familiarity with it, we provide a brief review of some basic features and syntax here. For more information, there is an enormous amount of free literature available that can be found with internet searches. Probably the best place to look is on the MATLAB website http://www.mathworks.com/help/matlab/index.html. The book Numerical Methods for Engineers and Scientists by Gilat and Subramaniam also has a nice MATLAB introduction in their appendix.

A.1 Basic operations

The most basic way to use MATLAB is for simple calculations. The symbols used for scalar operation can be found in Table A.1.

Table A.1. MATLAB operations.

Operation	Symbol	Operation	Symbol
Addition	+	Division	/
Subtraction	−	Exponentiation	^
Multiplication	*		

Mathematical expressions can be typed directly into the command window. For example:

```
>> 7 + 6/2
ans =
     10
>> (7+6)/2 + 27^(1/3)
ans =
      9.5000
```

Numerical values can be assigned to variables, which can then be used in mathematical expressions and functions:

```
>> a = 11
a =
     11
>> B = 3;
>> C = (a - B) + 40 - a/B*10
C =
     11.3333
```

MATLAB also includes many built-in functions. These functions can take numbers, variables, or expressions of numbers and variables as arguments. Examples of common MATLAB functions:

- Square Root:

  ```
  >> sqrt(64)
  ans =
        8
  ```

- Exponential (e^x):

  ```
  >> exp(2)
  ans =
        7.3891
  ```

- Absolute value:

  ```
  >> abs(-23)
  ans =
        23
  ```

- Natural logarithm (Base e logarithm):

  ```
  >> log(1000)
  ans =
        6.9078
  ```

- Base 10 logarithm:

  ```
  >> log10(1000)
  ans =
        3.0000
  ```

- Sine of angle x (x in radians):

  ```
  >> sin(pi/3)
  ans =
        0.8660
  ```

Relational and logical operators can be used to compare numbers, and are given in Tables A.2 and A.3. When using logical operators, a nonzero number is considered true (1), and a zero is false (0).

Table A.2. Relational operators.

Operator	Description	Operator	Description
<	Less than	>=	Greater than or equal to
>	Greater than	==	Equal to
<=	Less than or equal to	~=	Not equal to

Table A.3. Logical operators.

Operator	Name	Description
&	AND	Operates on two operands. If both are true, yields true (1). Otherwise, yields false (0).
|	OR	Operates on two operands. If at least one is true, yields true (1). If both are false, yields false (0).
~	NOT	Operates on one operand. Yields the opposite of the operand.

```
>> 6 > 9
ans =
     0
>> 4 == 8
ans =
     0
>> 3&10
ans =
     1
>>  a = 6|0
a =
     1
>> ~1
ans =
     0
```

The default output format in MATLAB is 'short' with four decimal digits. However, this may be changed by the user using the `format` command. Several available formats are listed in Table A.4. Examples are given below.

```
>> format short
>> pi

ans =

    3.1416

>> format long
>> pi

ans =

    3.141592653589793

>> format shorte
>> pi

ans =

    3.1416e+00
```

Table A.4. Display Format

Command	Description
format short	Fixed point with four decimal digits for $0.001 \leq number \leq 1000$. Otherwise, display format for 'short e'.
format long	Fixed point with 16 decimal digits for $0.001 \leq number \leq 1000$. Otherwise, display format for 'long e'.
format shorte	Scientific notation with four decimal digits

A.2 Arrays

An array is the form utilized in MATLAB to store data. The simplest array (a vector) is a row or column of numbers. A matrix is a more complicated array, which puts numbers into rows and columns.

A vector is created in MATLAB by assigning the values of the elements of the vector. The values may be entered directly inside square brackets:

```
>>var_name = [number number . . . number]
```

In row vectors, entries are separated by a comma or a space. In a column vector, they are separated by semicolons. Vectors whose entries have equal spacing can be specified using colon notation via

```
>>variable_name = m:q:n
```

where m is the first entry, q is the spacing between entries, and n is the last entry. If q is omitted, it is assumed to be equal to one. Several examples of constructing vectors are given below.

```
>> y = [1 19 88 2000 ]
y =
      1     19    88    2000
>> p = [1; 4; 8]
p =
      1
      4
      8
>> x = 3:2:11
x =
      3     5     7     9     11
>> x = -2:1
x =
     -2    -1     0     1
```

A matrix can be created row by row, where the elements in each row are separated by commas or spaces. Note that all rows in a matrix must contain the same number of elements. Examples of creating matrices are:

```
>> A = [1 2 3; 4 5 6; 7 8 9]

A =

    1    2    3
    4    5    6
    7    8    9

>> a = 1; b = 2; c = 3;
>> B = [a b c; 2*b b+c c-a]

B =

    1    2    3
    4    5    2
```

Arrays and their elements may be addressed in several ways. For a vector named *ve*, $ve(k)$ is the element in the kth position (counting from 1). For a matrix named *ma*, $ma(k, p)$ is the element in the kth row and pth column. Single elements in arrays can be reassigned and changed, or used in mathematical expressions.

```
>> V = [1 8 4 2 5]

V =

    1    8    4    2    5

>> V(3)

ans =

    4

>> V(3)+V(4)*V(1)

ans =

    6

>> A = [1 2 3; 4 5 6; 7 8 9]

A =

    1    2    3
    4    5    6
    7    8    9
```

```
>> A(1,3)

ans =

    3

>> A(3,1)

ans =

    7
```

MATLAB allows you to not only address individual entries in matrices, but also sub-matrices. To do this, the row and column argument can contain a vector of indices (1 being the first element) or a colon ":" for the whole row or column. Using the construction "m:q:n" is very useful here. Note that the indices do not need to be sorted and will return the corresponding entries in the requested order. Some examples are below.

```
>> MAT = [3 11 6 5; 4 7 10 2; 13 9 0 8]
MAT =
    3    11    6    5
    4     7   10    2
   13     9    0    8
```

– The whole column of column 2 and 4:

```
>> MAT(:,[2,4])
ans =
   11    5
    7    2
    9    8
```

– Row 2, columns 1 to 3:

```
>> MAT(2,1:3)
ans =
    4    7   10
```

– Row 3, 1, and 3 again (in that order) and the whole column:

```
>> MAT([3,1,3],:)
ans =
   13    9    0    8
    3   11    6    5
   13    9    0    8
```

This advanced indexing can also be used for vectors (of course only having one argument):

```
>> v = [4 1 8 2 4 2 5]
v =
      4      1      8      2      4      2      5
>> u = v(3:7)
u =
      8      2      4      2      5
```

Several built-in functions for arrays are listed in Table A.5.

Table A.5. Built-in functions for arrays.

Command	Description
length(x)	Returns the number of elements in vector x
size(A)	Returns a row vector of two elements, $[m, n]$, m and n are the size $m \times n$ of A.
zeros(m,n)	Creates a zero matrix of size $m \times n$.
ones(m,n)	Creates a matrix of ones of size $m \times n$.
eye(n)	Creates an identity matrix of size $n \times n$.
mean(x)	Returns the mean value of all entries in a vector x
sum(x)	Returns the sum of the entries in a vector x
sort(x)	Arranges the elements of a vector x in ascending order
det(A)	Returns the determinant of a square matrix A
A'	Returns the transpose of the matrix or vector A

A.3 Operating on arrays

Addition, subtraction, and multiplication of arrays follow the rules of linear algebra. When a scalar is added to or subtracted from an array, it is applied to all elements of the array.

```
>> u = [1 2 3]; v=[4 5 6]

v =

      4      5      6

>> w = u+v

w =

      5      7      9
```

```
>> u*v
??? Error using ==> mtimes
Inner matrix dimensions must agree.

>> u*v'

ans =

    32

>> u'*v

ans =

     4     5     6
     8    10    12
    12    15    18

>> A = [1 2;3 4;5 6]

A =

     1     2
     3     4
     5     6

>> B = [7 8;9 10]

B =

     7     8
     9    10

>> A*B

ans =

    25    28
    57    64
    89   100
```

Element-by-element multiplication, division, and exponentiation of two arrays of the same size can be carried out by adding "." before the operator. Examples of element-by-element operations:

```
>> A = [1 2 3;4 5 6]

A =

       1       2       3
       4       5       6

>> B = [7 8 9; 10 11 12]

B =

       7       8       9
      10      11      12

>> A.*B

ans =

       7      16      27
      40      55      72

>> A./B

ans =

   1.4286e-01   2.5000e-01   3.3333e-01
   4.0000e-01   4.5455e-01   5.0000e-01

>> A.^3

ans =

       1       8      27
      64     125     216
```

A.4 Script files

A script file (or program) is a file that contains a series of commands, executed in order when the program is run. If output is generated by these commands, it appears in the command window. A script file can be thought of as nothing more than a series of commands that are entered in sequence on the command line.

Script files are edited in the Editor Window. To make a new script file, select "New", followed by "M-file" from the "File" menu. A script file must be saved before it can be executed. Once saved, it can be run by entering the file name into the Command Window, or directly from the Editor Window.

Variables utilized in a script file can be set and changed in the file itself, or in the Command Window, either before the program is run, or as input during execution.

A.5 Function files

A function file is a program in MATLAB that is used a function. It is edited and created
in the Editor Window like a script file, but its first line must be of the following form:

```
function [outputArguments] = function_name(inputArguments)
```

-   `function` must be the first word in the function file.
-   If there are more than one input or output argument, they are separated by com-
    mas.
-   All variables in a function file are local; they are defined and recognized only in-
    side the file.
-   Closing the function with `end` as the last instruction in the file is optional.

The following is a simple example of a function:

```
function [y] = myfun(x)
    y = exp(x) - x^2;
end
```

This function returns $e^x - x^2$ for any given x. It can be called by

```
>> myfun(1)

ans =

    1.7183
```

You can turn the function defined as above into a handle that you can pass to a different function (this will be important in some of the chapters) with the expression
`@myfun`.

A.5.1 Inline functions

It is sometimes useful to define a function "inline" (not in a separate file as explained
above), if the function is a mathematical formula (basically if you can write it in a
single expression). The following code creates a function $f(x, y) = x + 2y + 1$ with two
arguments and assigns it to the handle `myfunction`:

```
>> myfunction = @(x,y) x+2*y+1;
```

and you can evaluate the function like this:

```
>> myfunction(2,0)
    3
```

A third option is to create an inline function like this:

```
>> myfunction2 = inline('sin(x)-exp(x)');
```

Here MATLAB automatically detects that x is the argument of the function. This does not work reliably with more than one argument. This is why we prefer the syntax above.

A.5.2 Passing functions to other functions

We will need to pass functions to other functions many times in this book. An example would be a function `integrate(f,a,b)` that computes (approximates) $\int_a^b f(x)$. You have two options for this:
1. Define the function inline:

```
>> myfunction = @(x) exp(x)-x*cos(x);
>> integrate(myfunction, a, b)
```

2. Or by creating a file `myfun1.m` with the contents

```
function y = myfun1(x)
y = exp(x)-x*cos(x);
end
```

    and then call

```
>> integrate(@myfun1, a, b)
```

Note that the @ symbol is required to turn the function defined in the .m file into a handle.

A.6 Outputting information

MATLAB has several commands used to display output, including `disp` and `fprintf`. The `disp` command accepts variable names or text as string as input:

```
disp(x)
disp('text as string')
```

The `fprintf` command has the form:

```
fprintf('text as string \%5.2f additional text\n', x)
```

The "%5.2f" specifies where the number should go, and allows the user to specify the format of the number.

A.7 Programming in MATLAB

In programming in MATLAB, some problems will require more advanced sequences of commands. Conditional statements allow for portions of code to be skipped in some situations, and loops allow for portions of code to be repeated.

A conditional statement is a command which MATLAB decides whether execute or skip, based on an expressed condition. The "if-end" is the simplest conditional statement. When the program reaches the "if" statement, it executes the commands that follow, if the condition specified is true. It then follows these commands down to the "end" statement, and continues executing the program. If the conditional statement is false, it skips the commands between "if" and "end" and continues with commands after the "end" statement.

Another conditional structure is "if-else-end". This provides MATLAB the means to choose one of two groups of commands to execute. If the condition associated with the "if" statement is satisfied, that code is executed, and the program skips to the code after the conditional structure (skipping the "else" section). If the it is not satisfied, the program skips this section of code and instead executes the commands in the "else" section.

If it is necessary for MATLAB to choose one of three or more groups of commands to execute, a similar structure is followed, with "elseif" statements added between the first "if" section and the last "else" section of commands.

Loops are another way to alter the flow of commands in a MATLAB program. Loops allow a group of commands to be repeated. Each repetition of code inside a loop is often referred to as a pass.

The most common type of loop in MATLAB is the "for-end" loop. These loops dictate the the group of commands inside the loop are run for a predetermined number of passes. The form for such a loop is:

```
for k = f:s:t
    ...
    ...
end
```

In the first pass, k = f. After the commands up until "end" are executed, the program goes back to the "for" and the value for k is changed to k + s. More passes are done until k=t is reached. Then, the program goes to the code past the "end" statement. For example, if k = 2:3:14, the value of k in the passes is 2, 5, 8, 11, and 14. If the increment value s is omitted, the default value is 1. Note that any vector can be given to the right of the equal sign and the loop will iterate over every element of the vector.

A.8 Plotting

MATLAB has a variety of options for making different types of plots. The simplest
way to make a two-dimensional plot is using the plot command: "plot(x,y)". The arguments x and y are each vectors of the same length. The x values will appear on
the horizontal axis and the y values on the vertical axis. A curve is created automatically connecting the points. The following code is an example of a basic plot, which
is shown in Figure A.1.

```
>> x = [1 2 4 5 6 6.5 8 10];
>> y = [2 7 8 5.5 4 7 9 8];
>> plot(x,y,'.-')
```

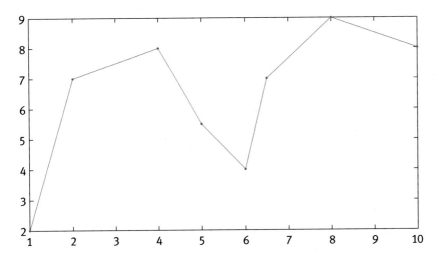

Fig. A.1. A plot of data points.

With the "plot" command, there are additional arguments which can be used to specify line color and marker type, if markers are desired. The command takes the form
"plot(x,y,'line specifiers')". A list of there arguments is given in Table A.6.

Table A.6. Line and marker specifers.

Line Style	Specifier	Line Style	Specifier
solid (default)	–	dotted	:
dashed	–	dash-dot	-.

Line Color	Specifier	Line Color	Specifier
red	r	blue	b
magenta	m	black	k
green	g	cyan	c
yellow	y	white	w

Marker	Specifier	Marker	Specifier	Marker	Specifier
plus sign	+	asterisk	*	square	s
circle	o	point	.	diamond	d

A.9 Exercises

1. Create a vector of the even whole numbers between 29 and 73.
2. Create a column vector with 11 equally spaced elements, whose first element is 2 and whose last is 32.
3. Let x = [2 5 1 6].
   (a) Add 16 to each element.
   (b) Add 3 to just the odd-index elements.
   (c) Compute the square root of each element.
   (d) Compute the square of each element.
4. Given x = [3 1 5 7 9 2 6], explain what the following commands "mean" by summarizing the net result of the command.
   (a) x(3)
   (b) x(1:7)
   (c) x(1:end)
   (d) x(1:end-1)
   (e) x(6:-2:1)
   (f) x([1 6 2 1 1])
   (g) sum(x)
5. Given the arrays x = [1 4 8], y = [2 1 5] and A = [3 1 6 ; 5 2 7], determine which of the following statements will correctly execute and provide a result. If the command will not correctly execute, state why it will not.
   (a) x + y
   (b) x + A
   (c) x' + y

(d) `A - [ x' y' ]`

(e) `[ x ; y' ]`

(f) `[ x ; y ]`

(g) `A - 3`

6. Given `x = [1 5 2 8 9 0 1]` and `y = [5 2 2 6 0 0 2]`, execute and explain the results of the following commands:

(a) `x > y`

(b) `y < x`

(c) `x == y`

(d) `x <= y`

(e) `y >= x`

(f) `x | y`

(g) `x & y`

(h) `x & (~y)`

(i) `(x > y) | (y < x)`

(j) `(x > y) & (y < x)`

7. The exercises here show the techniques of logical-indexing (indexing with 0-1 vectors). Given `x = 1:10` and `y = [3 1 5 6 8 2 9 4 7 0]`, execute and interpret the results of the following commands:

(a) `(x > 3) & (x < 8)`

(b) `x(x > 5)`

(c) `y(x <= 4)`

(d) `x( (x < 2) | (x >= 8) )`

(e) `y( (x >= 3) & (x < 6) )`

8. Given `x = [3 15 9 12 -1 0 -12 9 6 1]`, provide the command(s) that will

(a) set the values of x that are negative to one;

(b) multiply the values of x that are even by 5;

(c) extract the values of x that are greater than 10 into a vector called y;

(d) set the values in x that are less than the mean to zero.

9. For the function $y = \frac{x^3+1}{(x^2+2)^2}$, calculate the value of y for the following values of x: $-2.65, -1.85, -1.02, -0.21, 0.61, 1.3, 2.4, 3.9$. Make a plot of the points with the asterisk marker for the points and a red line connecting them.

10. Show analytically (i.e. without a computer) that the infinite series $\sum_{n=0}^{\infty}(-1)^n\frac{1}{(2n+1)}$ converges to $\frac{\pi}{4}$. Compute the sum of the series using a for loop with $n = 100$, $n = 1000$, $n = 10000$, and $n = 100{,}000$. Does it converge to the correct solution?

11. In the MATLAB/Octave programming language, for loops are slow, and it is generally much faster to do the same operations with vectors when possible. Repeat the last problem and time (using tic and toc) the for loop. Next, do the same calculations but with vectors, by making a vector whose first term is 0, and the rest of the terms are $(-1)^n\frac{1}{(2n+1)}$, and use the sum function to add the terms. Time the vector operations. Which is faster?

12. Create a plot in MATLAB that plots the function $f(x) = \cos(2x)$ and $g(x) = \sin(x)$ into *one* plot for x in $[0, 2\pi]$. Pick two different line styles (your choice) and include a legend.

13. Write a function `even_numbers(x)` that given a vector of natural numbers returns a vector that only contains the even numbers. Example:

```
>> even_numbers([2 3 4 4 1 2222 2])
    ans = [2 4 4 2222 2]
```

Can you do this without manually looping over the input vector? (you can add an element z to an existing vector y using `y=[y z];` if you need it)

14. Write a function `x=fiddle(a,b,n)` that returns the value $x = F(a, b, n)$ based on the following recursive definition:

$$F(a, b, 0) = a, \qquad F(a, b, 1) = b, \qquad F(a, b, n) = F(a, b, n-1) + F(a, b, n-2).$$

Check that it produces the right answer by outputting `fiddle(1,3,n)` for $n = 1, \ldots, 10$ with your function.

Hint: A function can call itself (with different parameters). Make use of that.

Made in the USA
Columbia, SC
26 August 2020

17606987R00084